THE DISSIDENT VOICE

THE DISSIDENT VOICE

POEMS OF N.M. RÁSHED

Translated from Urdu
by
M.A.R. HABIB

MADRAS
OXFORD UNIVERSITY PRESS
DELHI BOMBAY CALCUTTA
1991

Oxford University Press, Walton Street, Oxford OX2 6DP
New York Toronto
Delhi Bombay Calcutta Madras Karachi
Petaling Jaya Singapore Hong Kong Tokyo
Nairobi Dar es Salaam
Melbourne Auckland
and associates in
Berlin Ibadan

SBN 0 19 562741 5

Typeset by Jayigee Enterprises, Kamala Arcade, Madras 600 006
and printed by Vivek Sachdev
at India Offset, Madras 600 014
and Published by S.K. Mookerjee
Oxford University Press, 219 Anna Salai, Madras 600 006.

For my mother and father

Acknowledgements

I should like to thank Mrs Sheila Ráshed for her kind permission to look at her husband's unpublished manuscripts and for her support of this project. I am greatly indebted to Dr David Matthews of the School of Oriental and African Studies in London both for his valuable time and helpful advice. A number of people kindly agreed to read through my script: Terry Eagleton, Ervene Gulley, Chris Fitter, Chris Clare and Tracy Reed all offered insightful suggestions for revision. Over the past two years or so I have profited from discussions with the Urdu poets Iftikhar Arif, Saqi Farooqi and many others.

My greatest single debt is to the Urdu poet Akbar Hyderabadi who gave me the benefit of his expertise in going through these translations with me line by line. As for my parents' help and support, it suffices to say that without them this project would have been impossible. Finally, I must acknowledge that any infelicities or inaccuracies of translation are entirely my responsibility.

Preface

As with any language, Urdu exhibits its own peculiar forms of resistance to translation. But, in a sense, the word 'translation', as applied to any literature, is a misnomer. Quite apart from the impossibility of reproducing in English the powerful musical cadences of the Urdu language, the translator today faces a more complex acknowledgement. Much modern European philosophy and literary theory has insisted that 'meaning' is not a stable independent entity but rather something that is bound to a particular context. Modernist literature begins with the premise that even our own language cannot adequately express the world of our experience. This insight governs the form, and sometimes the content, of the work of James Joyce, Virginia Woolf, Ezra Pound, T.S. Eliot and Samuel Beckett, to name but a few writers. What hope, then, can there be for the translator? If language is intrinsically problematic, the very act of conceptualising experience, of imposing on the original unity of our experience an artificial distinction of subject and predicate, is to falsify it. There is no 'natural' correspondence between a word or sign and what it signifies. How much more difficult it must be, then, to translate: to substitute one series of falsifications for another, to transplant the very seeds of inadequacy to an alien soil. If we accept, as we must, that what we call experience is neither a homogeneous entity discoverable by refining our methodologies nor something which can be predefined abstractly by philosophy, then we must move to a perception of it as defined historically, socially and politically.

A single phrase, then, may demand location within an entire cultural context to yield its meaning. How otherwise, for example, could we understand the weight of King Oedipus's insulting words to the blind prophet Tiresias in *Oedipus Rex*? Or the beginning of Homer's *Iliad* or of Dante's *Inferno*? Yet what we do effectively in translating is to bring forth the phrase wrenched from its context: we pretend that this transfer to a new context somehow leaves the phrase unaffected in its significance. For the most part we concentrate on the 'minute particulars' of the possible grammatical parallelisms between one language and another, substituting an equivalent when the original words are meaningless in their displaced home. Viewed in this way, translation is by nature a doomed endeavour. But once we grant its inescapable subjectivity, we can see that translation is no more futile than the process of reading itself. It's a common enough notion, recognised by Wordsworth, that every act of reading is an interpretation, a rewriting. Equally, every act of translation is an interpretation, reflecting as much of the translator and his or her background as it does of its author. Thus Pope's Homer is a very different poet from Lattimore's Homer; Burton Raffel's Horace is a different man altogether from Milton's Horace. These differing versions stand beckoning us towards the original but the 'original' is no more than the path lit up by these versions. There is no terminus at which we can gaze on the face of the original in its uninterpreted, unveiled purity.

Traditionally, a number of strategies have been adopted to minimise the loss of meaning in translation. One of these is simply to translate into a familiar idiom. But what do we gain by this? One translation of the *Iliad* uses modern English and is replete with modern colloquialisms. This makes the Homeric heroes accessible to us, portraying them in the language we use to portray our own world. But don't we also lose by this procedure? For one thing, it relies too heavily on the power of 'ordinary' language, a notion deeply embedded in illusion. Surely most of the best poetry, especially over the last two centuries, has achieved the status of poetry precisely because of its departure from conventional language which embodies conventional perceptions of the world. More importantly, this familiarising use of language utterly destroys the real differences between the worlds of author and translator:

it assumes, for example, that the Homeric gods are part of our world and can be spoken of in the same language. In a very broad historical sense, that is true. But such a breadth of meaning would make translation irrelevant: those gods are part of our world just as knowledge of ancient Greek is part of the historical awareness informing our present. Surely one function of translation must be to express the very remoteness from our culture of the Homeric gods, myths and ethics.

A guiding principle of the present work is that a poem, even in translation, should retain something of its foreignness, its distant flavour. That's not to say that poems should be translated literally: the myth of literal meaning has also been exploded in our time. That would again presuppose that one primordial meaning acts as the subtratum on which other variant meanings rest. But translation might profitably move somewhere between these two hypothetical extremes of literal meaning and complete familiarisation. The language could be familiar enough to read coherently in the translated version (though incoherence is itself a vehicle of modern poetry) but not so familiar as to read like a poem written by an English poet. To fall into the latter extreme would be to deprive the poem of its status as a bearer and mirror of its own culture.

A look at one or two of the current translations may clarify the assumptions behind them. In the poem 'Near the Window' I have rendered the first line as: 'Lamp of Love's chamber, awake!' A more 'literal' rendering would read: 'Awake, lamp of the chamber of union!' and a familiarising version might be: 'Wake up, winsome bedroom's light;'.[1] The problem with the 'literal' version is that it's unintelligible to an English or American reader, unaware of the primarily romantic (and mystical) significance of the word 'union' in Urdu and Persian poetry. On the other hand, the more colloquial reading destroys both the metre of the original as well as its deliberate archaism: the diction of the original line is Persianised and

---

[1] This is the translation employed by Carlo Coppola and M.H.K. Qureshi, *The Beloit Poetry Journal* (Winter 1962–3), pp. 48–50.

locates itself self-consciously within a certain tradition of love poetry. Only against this stylised opening can the original poem's own subsequent colloquialism, signalling its departure from that tradition, effectively flower. My translation attempts a compromise: while it substitutes the word 'love' for 'union', it retains an alliteration ('lamp' and 'love') which also punctuates the original; it seeks to preserve something of the original metrical flow by commencing with a trochee (a stressed syllable followed by an unstressed one) rather than an iamb or a spondee. Moreover 'Love' is capitalised as if to express the general tradition of love in which this poem's love is situated, rather than referring initially to a specific love. Whether or not this attempt succeeds is of course for the reader to judge.

In Urdu, certain refrains and repetitions can be effective, intensifying an emotional impact through their powerful rhythms. But on occasion, these can prove lame in English. For example, the first stanza of 'Deserted Sheba' reads:

> Solomon, head on his knees and Sheba desolate
> Sheba desolate, Sheba the home of ghosts
> Sheba an abysmal heap of sorrows
> A land devoid of grass, verdure and flowers

In English the phrase 'head on his knees' is somewhat cumbersome and might be replaced with 'head in his hands' which both retains the mood and equally sustains the original metre. Again, the repetition of 'Sheba', although effective here, might be reduced. The Urdu word for 'heap' also means 'reservoir'. 'Abysmal heap' or even 'endless heap' is somewhat incongruous as a visual image. Playing on the dual meanings the Urdu word, I have rendered the phrase as 'abysmal lake'; this introduces Miltonic connotations, suggesting that Sheba is a type of hell. True, 'lake' is inconsistent with the overall atmosphere of aridity in the poem but Milton's lake in *Paradise Lost*, we remember, is a lake of fire. I have chosen to render 'land' as 'world' so as to induce a perception of Sheba as an isolated place, unique in its estrangement. Hence my version reads:

> Solomon, head in his hands, and Sheba desolate
> Sheba desolate, a home of ghosts
> Sheba an abysmal lake of woes
> World devoid of grass, greenery and flower

Another major problem is that some forms of Urdu poetry have traditionally been discrete, lacking the unity which Western readers customarily expect from a poem. In translating some of Ráshed's longer poems, I have occasionally tried to follow up a line of imagery or metaphor a little more consistently than it stands in the Urdu, without attempting to impose an artificial unity on the entire poem. At times I have used singular forms instead of plurals and vice versa; simple past forms rather than the long-winded and sometimes obtrusive past perfect continuous; and active rather than passive forms except where the latter are integral to meaning. I have tried as far as possible to indicate in the footnotes occasions where I have diverged substantially from the original grammar. I should return, finally, to my initial point: these translations are offered as one of many possible interpretations of Ráshed's work. They present no claim to finality but I hope that they reveal, at worst, the subtleties of Ráshed's sensibility and, at best, a glimpse into his undoubted poetic stature.

Contents

From *X = Man* [*LA MUSAWE INSAN*]

From *The Possibility of Doubt* [*GUMAN KA MUMKIN*]

Introduction

It is one of the saddest ironies of history that two religions with much in common, Islam and Christianity, have been at odds for centuries. One might be forgiven for thinking that in the twentieth century the prevailing image of Islam stamped on Western minds is that of a veiled woman with a machine gun. But these associations of violence, terrorism and fanaticism, acquired by Islam in the Western imagination, have helped blind us to the memory of the splendid literary and cultural heritage of that religion. The voice of Islam as heard in the West is but a faint and falsifying echo of what once were a myriad tongues celebrating the full range and depth of human experience. Their history still lives and their song can still be heard, if we listen.

Most of us are familiar with Omar Khayyam's quatrains; not so many are aware of the risks taken by this thinker in opposing religious orthodoxy. And those who imagine that Islam is a religion which cannot brook criticism might do well to explore the vast tradition of self-questioning and revaluation in Islamic literature and philosophy, a tradition which runs from Sufi mystical philosophy through Khayyam to the modernist poetry of N.M. Ráshed, whose verse is presented here for the first time in English translation.[1] As early as the ninth century, the Arab poet Abu Nawas wrote poems devoted to pleasure in open defiance of religious strictures. The Arab thinker Ibn Hazm wrote, some eight hundred years before Freud, a profound analysis of love and the hidden motives

[1] A few of Ráshed's poems have been translated into English, most notably in *The Penguin Book of Modern Urdu Poetry*, tr. Mahmood Jamal (Middlesex: Penguin, 1986). Some poems have also been translated into French.

behind human actions. Again, long before Descartes took to doubting himself, the eleventh century Islamic philosopher al-Ghazzali used 'doubt' as an epistemological tool. The veil, to some extent, may be over our own eyes; and beneath her veil the woman may be beautiful in ways we had not imagined. It is within this tradition of religious and political heterodoxy that Ráshed's poetry took root.

Ráshed's Life and Literary Career

Aristotle's dictum that 'Man is a political animal' finds ample exemplification in the life and work of Nazar Muhammad Ráshed.[2] Born on 9 November 1910 in the obscure market town of Akalgarh in West Pakistan, his social and public persona was rapidly nurtured on the soil of his early poetic interest. Within the next forty years or so he came to be recognised, along with Faiz Ahmed Faiz and Miraji, as one of the three major modern poets of the Urdu language.

Urdu, like English, is an Indo-European language. Just as English combines elements of ancient Greek, Latin, Anglo-Saxon and French, Urdu is a hybrid of Persian, Arabic and dialects of northern India. But Urdu is written in the Arabic script, from right to left. Its poetry, moreover, is composed in quantitative metre (according to syllable length rather than stress) like Latin and Greek verse. 'Urdu', originally meaning 'camp', derives from the same Turkish root as the English word 'horde'. It arose in the twelfth century through the contact of north Indian idioms with the Arabicised Persian and Turkish of the Turko-Mongolian and Afghan Muslim conquerors arriving in north west India. It was the language spoken in the camps of these invaders, hence its name. Urdu today is the national language of Pakistan as well as one of the official languages of India. It has been spread by emigration to many other countries where there exists a

[2] This account of Ráshed's life is based on a number of résumés as well as two manuscripts, one in English and one in Urdu, written by Ráshed himself. These documents were kindly made available to me by Mrs Sheila Ráshed.

substantial Muslim community such as Britain and parts of the United States. Ráshed's own poetry, richly absorbative of Arabic, Persian and Western influences, bears ample testimony to the varied and sometimes troubled history of his language. Take this poem, for example, which begins with a heavily Persianised diction:

> Lamp of Love's chamber, awake!
> Wake from this joyful floor of soft dreams...

But from here the poem moves to an unadorned and somewhat irreligious evocation of sordid city life:

> Beneath this minaret's shadow, I remember
> A mournful, penniless priest
> Drowsing in a dark, hidden corner
> Like a useless god...

Ráshed's earliest poems were written in traditional forms. It was around 1932 that he started writing free verse in Urdu. This represents a lag of some twenty years behind European poets. Ráshed also published critical essays and was later to translate T.S. Eliot's famous essay 'The Three Voices of Poetry'.[3] Ráshed produced four volumes of poetry between 1942 and his death in 1976. He travelled widely, living for long periods in Eastern and Western countries, an experiential synthesis reflected in his verse. He was influenced both by Arab and Persian writers (such as al-M'arri and Rumi) as well as the French symbolists and Western modernists in the vein of T.S. Eliot and Ezra Pound.

Of the three outstanding poets in modern Urdu literature, Faiz, whose verse is relatively accessible, has achieved the most popularity. But Ráshed's poetry is obscure, allusive and questions the very nature of language. And, as with Eliot's verse, these qualities conspire to deny it any 'popular' appeal. Those who respect and enjoy Ráshed are primarily intellectuals

[3] This translation, published in the journal *Nai Tahreeren* in September 1954, is cited in Donald Gallup, *T.S. Eliot: A Bibliography* (London: Faber, 1969), p. 339.

prepared to welcome novelty and difficulty. Moreover, Ráshed sometimes expressed towards religion a cynicism so deep and cutting that he did wound and alienate the orthodox. But a religious dismissal of Ráshed does a disservice to him and overrides some important distinctions. Ráshed possessed a profound knowledge of Islamic culture, encompassing Arabic, Persian, Urdu and Punjabi.

Ráshed's poems are a historically accurate expression of the problems facing the Indian subcontinent and the Islamic world from the period between the two World Wars to the present. Islam, like any other religion, has existed both through doctrine and through cultural tradition. Ráshed took exception not to the doctrines of Islam *per se* but to the particular fatalistic, dogmatic and inert interpretation of Islam he saw as pervading the cultural practices of the subcontinent: it was this passive spirit of submission, he felt, that had contributed to Asia's decline in the modern world. But it is often assumed, quite incorrectly, that he was an atheist. Although Ráshed was cynical towards orthodox religion, his near obsession with religion reflects his spiritual struggles and his failure to find an effective religious solution. Even T.S. Eliot, a confirmed Anglo-Catholic, had said that in the modern world religious doubt is 'a variety of religious belief': it is part of the process whereby one arrives at God. For all his rebellion against dogma and rigidity of custom, Ráshed's sensibility was inevitably shaped by religion, and the bitterness of his disillusionment with his spiritual heritage betrays a continuing dialogue with his sense of the Divine. This is borne out by the comments of his widow, Sheila Ráshed, who states in a letter: 'My husband was not what one would call a practising Moslem. But he was a deeply religious person in the true sense of the word'.[4]

Ràshed's foiled religious search ended, in fact, in a broad humanism, encompassing a hatred of totalitarianism and a belief in the sacredness of individual rights.[5] In Ráshed's own

[4] Sheila Ráshed, letter addressed to Mr Shahid, 3 September 1976, p. 3.

[5] Sheila Ráshed, letter to Mr Shahid, p. 4.

words, he tried in his verse 'to stress some of the current problems, the foremost among them for me being the alien rule, religious dogmatism, moral repressions... which had continuously dwarfed the Asian soul...'[6] Ráshed's credo seems, at first sight, to be an unabashed bourgeois liberal individualism, which hardly sounds revolutionary today. But to understand why Ráshed's political, religious and aesthetic attitudes were iconoclastic, we must go a little more deeply into his context.

Classical Urdu Poetry and Its Background

In talking of the Urdu literary traditions from which Ráshed sprung, we need to consider two factors which loom large in the connection between the development of Urdu poetry and its social setting. To begin with, Urdu literature was largely, though not exclusively, the cultural expression of the Indian Muslim community, to whose fate it was closely tied. Under Mughal rule the language of court and state was Persian, which was patronised as the language of literary expression among Muslim circles. Though Urdu had developed in the north, it was first cultivated as a literary language in the Deccan area of south India, partly because the Muslim rulers of these southern kingdoms patronised the writing of Urdu alongside Persian. The early eighteenth century marks the beginning of Classical Urdu poetry, which flourished in Delhi and Lucknow. These poets included Sauda, Dard, Mir, Zauq, Momin and Ghalib (d.1869), who is generally acknowledged as the greatest Urdu poet.

The second factor determining poetic development was that, under Mughal administration, society was in many ways static. It was a feudal structure where the king and nobility provided for the intellectual and cultural life of the society. In marked contrast to Europe where Protestantism often

---

[6] 'Modern Urdu Poetry: The text of an Interview with Mr N.M. Ráshed, by Carlo Coppola, of Chicago University,' MS., p. 2. Hereafter cited as Coppola.

impelled social change, the most fundamental feature underlying conservatism and fatalism of outlook was religion. It is against the backcloth of this heritage that Ráshed's reaction against religion must be viewed. Islam is not by nature a conservative religion but it had become embedded in the prevailing social structure in a way that might bear comparison with pre-Reformation Europe where the overarching world-view was that of Roman Catholicism. The social emphasis on authority and tradition was reflected to some extent in Classical Urdu poetry, perhaps more in form than in content. The entire body of traditional Urdu poetics, including prosody, poetic form, ideas of diction and subject matter, was in fact taken over almost exclusively from Persian.

By far the most characteristic poetic form was the *ghazal*, which can be traced back to its Arabic source as written by Abu Nawas (d. 813). Ghazal derives from an Arabic word meaning 'conversation with one's beloved'. By the time the ghazal was taken over into Urdu, it had developed certain strict conventions, both formal and thematic. Traditionally, it was a short but metrically sophisticated love poem of up to thirty distichs. The first and last distichs are couplets, called *matla* and *maqta* respectively, and the overall rhyme scheme is *aa ba ca da ea fa aa*. The ghazal's intricate form and content perhaps reflect two striking features of Mughal society. Until the latter part of the nineteenth century, ghazal writing was closely associated with courtly and noble circles. This affected the whole concept of innovation in Urdu poetry. Various Urdu scholars have pointed out that what makes for greatness in Classical Urdu poetry is not the introduction of striking new images or forms or areas of experience but rather the ingenious and novel juxtaposition of traditional elements.[7] There are several parallels to this in Western literature. One might think of the Homeric epics where the notion of individual authorship, as we know it, is absent. Whoever composed

[7] See, for example, D.J. Matthews and C. Shackle, *An Anthology of Classical Urdu Love Lyrics: Text and Translations* (London: Oxford University Press, 1972), pp. 10–11.

these was constrained by the conventions of an oral epic tradition which included prescribed formulae, epithets, repetitions, catalogues and other narrative devices. The narratives were already communal property which an 'individual' poet could merely modify. As late as the eighteenth century, Alexander Pope was still adhering very closely to these conventions in 'The Rape of the Lock', although in a somewhat more self-conscious and ironic manner. In the same way, 'individuality' in Classical Urdu poetry was not coterminous with our modern bourgeois understanding of it. It represented, rather, the freedom to exercise choice *within* a constraining framework of set themes and values. A poet could be innovative within a ghazal but was not at liberty to challenge the form of the ghazal or its enabling circumstances in either social or literary tradition. Given that this tradition of ghazal writing has persisted into our own century along with vestiges of its social determinants, it becomes easier to understand how Rashed's fierce defence of individualism can be viewed as subversive. In the West, it was only with increasing bourgeois political ascendancy, enshrined in bitter revolutions from 1789 to 1870, that a complete literary individualism found expression in the literature of the Romantics. The middle classes in India never really became a cohesive and influential group until the later nineteenth century.

A Western reader, bred on notions of 'organic unity' and coming to the ghazal for the first time, will perhaps be struck by the discreteness, the apparent mutual isolation in thought and mood, of each distich. But the ghazal was composed primarily for recitation at a gathering of poets known as a *mushaira*, a tradition which flourishes to this day. The audience would show its appreciation of each couplet or distich in turn, and to hold its attention, variety and verbal skill stood the poet in better stead than an overall unity of the poem which would have demanded prolonged study. Again, variation on familiar, stock themes would make for more immediate impact than the presentation of entirely original matter.

The second social feature influencing the ghazal was the status of love in Mughal society. As the Urdu scholar Ralph Russell has argued, the love treated of in the ghazal is essentially illicit love and it is always tragic insofar as it is unfulfilled. Moreover, 'the lovers were held to be violating laws upon which the very foundations of ordered social life were based'.[8] One can see how, in a society where romantic 'love' is still to some extent proscribed, Ráshed's emphasis on sensual love would be considered disruptive. The significance of the ghazal's content, the tragic nature of love, might become clearer if we remember that in Western society also the status of women has long been belittled. Against such a history, several English texts, such as *Wuthering Heights*, amply bear out the subversive nature of love, a powerful and frightening force which disrupts social and family ties. A more renowned example is of course Shakespeare's *Romeo and Juliet* where love is defined by its very privacy and covertness, being sharply opposed to social bonds and family loyalties. Shakespeare even makes the two lovers speak a language more poetic and refined (sometimes in sonnet form)[9] than the bawdy and vulgar idiom of social gaiety. In both English and Urdu texts the lovers themselves are aware of the disruptive force of their emotions and this contributes to their tragedy.

In contrast to English, literary Urdu is an intrinsically symbolic language which was until very recent times sharply differentiated from the everyday idiom. In a sense, of course, every language is symbolic and, modern theorists would argue, metaphorical, but the difference in question is one of degree, as determined by literary tradition and consensus. When the sixteenth century English poet Spenser wrote the *The Faerie Queene*, he was able to draw on certain widespread symbols such as the Christian cross, the shield, the broad highway to Hell, as well as other well-known images from the Bible such as the whore of Babylon. But with these symbols Spenser painstakingly constructed an intricate allegory:

---

[8] Ralph Russell and Khurshidul Islam, *Three Mughal Poets: Mir, Sauda, Mir Hasan* (London: George Allen & Unwin, 1969), pp. 98–9.

[9] See I, v. 92–108.

the allegory was not built into the language at his disposal. Indeed, when Coleridge insisted that poetic language was *intrinsically* symbolic, he was obliged to defend his poetic practice with a sophisticated account of the poetic imagination as a special faculty which destroyed and created anew the reality designated by conventional relations between words.[10] Again, through Shelley's words, the West Wind became a symbol of certain subjectively projected emotions and needs, just as for Keats the nightingale's significance transcended its usual associations. For these poets, *every* set of related words was symbolic, and the newly created symbols replaced the conventional ones. The Urdu language moves somewhere between these two extremes of symbolism as found in Spenser and the Romantics: there are so many symbols in literary Urdu that the language is already allegorical. On the other hand, Urdu poets have traditionally aimed not to undermine these symbols, as the English Romantics did, but to extend and refine them. Even if a poet opposed the religious or political status quo, he would conduct his argument on the received symbolic level; it was left for modern Urdu poets to deconstruct the symbols themselves.

The nature of allegory is to have many levels of meaning, and symbols are a constitutive element of this plurality. But in Urdu this openness is achieved somewhat differently from English. Urdu poets have long had a tendency towards abstractness: in the ghazal, the object of love can be any number of different things: the betrothed or wife of another man, a male youth or a courtesan. Like homosexuality, resort to courtesans was formally frowned upon, but it was a part of the society in which poets and their patrons moved. In each of these situations the lover knew that his love must ultimately be hopeless, hence some of the stock characters in the ghazal were: the wretched lover (*ashiq*), the merciless beloved (*mashuq*), the lover's lustful rival (*raqib*), and the cup-bearer at the wine party (*saqi*).

---

[10] See Samuel Taylor Coleridge, *Biographia Literaria* (London: J.M. Dent, 1975), pp. 167, 174.

These essentially formal features are augmented by another characteristic deriving from the ghazal's Persian heritage, the use of Sufi symbolism. In his study of Sufism, A.J. Arberry gives a list of symbols commonly used in Persian mystical verse: *zulf* (tress), referring to the revealed omnipotent attributes of God; *sharab* (wine), representing ecstatic experience at the revelation of the beloved, destroying the foundations of reason; and *saqi* (wine bearer), which stands for Reality as loving to manifest itself through all its revealed forms.[11] These are just a few of a vast stock of symbols andRáshed makes use of a number of them. We can see from this list that the foundational ambiguity of the ghazal lies in the object of love, which can be either human or Divine.

Some of these characteristics are exhibited in the following ghazal by the greatest Urdu poet Ghalib (1797–1869) who wrote just prior to the impact of the West on Urdu letters:[12]

Union with our beloved was not in Fate's Design
Had life gone on, we'd still in endless waiting pine

We have lived on your promise but in disbelief
Had we trusted, would joy not be deadly (though divine)?

Who can see Him, for He is peerless and Unique?
Were there trace of Otherness, we'd have known its sign

These plights of Sufism, Ghalib! And this your account!
A saint we'd think you, were you not given to wine!

The meaning of the first couplet is very clear on one level. The word for 'union' here is *wisal* which refers to both sexual union with a lover and mystic union with God. The second line incorporates an illusion to the Sufi doctrine of *fana* which can be translated as 'dying to self', a condition of union with God. So the interpretation of this line can be expanded to mean that without a complete death, physical as well as spiritual, one cannot unite with God. Hence this becomes a statement of cynicism with regard to union with God while on earth.

---

[11] *Sufism: An Account of the Mystics of Islam* (London: George Allen & Unwin, 1963), pp. 113–114.

[12] All translations used here are mine, unless otherwise stated.

The second distich also brings out Ghalib's understanding of the contradictory nature of his faith both in God and his lover. The final couplet involves an additional play on words such as 'wine' in their earthly and Divine connotations. One can see here the immense possibilities of irony yielded by the ghazal's form.

Poetic individuality was achieved through form rather than content. But it is deeply ironical that the very form of the ghazal was subversive, almost by definition, on religious and social levels. Though wine and romantic love were strictly forbidden, Urdu and Persian verse abounds with the imagery of wine-cup, wine-bearer, parties and boasts about mistresses. This use of 'illicit' imagery was sustained and protected partly by the ghazal's intrinsic duality of reference to an earthly lover and God. But also, in common with the way that many societies have dealt with potentially explosive subversive elements, the ghazal provided a relatively harmless arena into which unconventional thoughts and feelings could be channelled and tested. In the twentieth century the ghazal, along with other Indian literary forms, became integrated into a more widespread political movement against British rule. In this context, the ghazal became a powerful political weapon.

Modern Movements in Urdu Literature

After Ghalib, Urdu letters began to feel the exposure to Western influence. Governing the advent of this influence was the replacement, in 1835, of Persian by English as the state language of India. This gave access, especially through educational institutions, to Western thought. Another factor was the decline of the aristocracy and the growth of the Muslim middle class in the later nineteenth century, a class for which Urdu literature began to cater. The modern movement in Urdu at first comprised the 'Aligarh School' of writers, inspired by the Muslim reformist Sayyed Ahmad Khan (d. 1898). Sayyed Ahmad helped to modernise the outlook of the previously conservative Muslim community. He attempted to reconcile Qur'anic teaching with what he considered to be a more Westernised, 'scientific' and 'rational' approach to knowledge.

The other main figures of the Aligarh movement included Hali (d. 1914) who, considered the founder of literary criticism in Urdu, insisted on poetry being experiential, realistic and a vehicle for social morality. Two further writers, Azad and Shibli, helped to establish the writing of biography and history respectively. This modernising movement culminates in Sir Muhammad Iqbal who is generally considered to be the greatest Urdu poet of the twentieth century. As the only modern writer in Urdu who can claim the title of philosopher, Iqbal and his influential ideas, which attempted to reconcile Eastern and Western thought, are worth trying to understand at some length. According to him, the decadent condition of Islam, and specifically that of the Indian Muslims, lay in the pantheistic and fatalistic elements in Islamic thought inherited from figures such as the Spanish mystic philosopher Ibn Arabi. Iqbal saw himself as a spokesman for the Islamic world and his philosophy, influenced by Nietzsche, Bergson and the Persian mystic Rumi, reacted against this fatalism and ascetic ethic of self-denial. It affirmed the reality of Self as its basis. In the introduction to a translation of his long Persian poem *Asrar e Qudi (Secrets of the Self)*, Iqbal states that the fundamental fact of the universe is what the English neo-Hegelian philosopher F.H. Bradley called the 'finite centre', or basic unit of experience.[13] Iqbal extrapolates this concept, somewhat against the grain of Bradley's philosophy and more in line with Bergson's, to interpret it as a centre of will and energy which gains mastery over its environment as it develops its individuality. The ego is fortified by *ishq* or love; this represents, for Iqbal, the desire to assimilate and absorb, its highest form being the creation of values and ideals, as well as the struggle to realise these. But, for Iqbal, the highest form of individuality is God, and the human individual becomes more

---

[13] Sir Muhammad Iqbal, *The Secrets of the Self*, tr. R.A. Nicholson (Lahore: Ashraf Publications, 1978) pp. xvii ff.

complete as he or she approaches nearer to God.[14] This type of movement is characteristic of most neo-Hegelian philosophies around the turn of this century.

Iqbal's world-view is not based on the crude opposition of 'science' (or 'reason') and faith that had characterised the 'modern' insights of thinkers either in the preceding Urdu tradition, such as Sayyed Ahmad, or those strands of nineteenth century Western thought, such as the positivism of Auguste Comte and Herbert Spencer, which had nourished themselves on such a dichotomy. Iqbal, on the contrary, undertook a modernistic appraisal of the nature of science. It did not represent for him a set of indubitable laws toward the certainty of which other disciplines, including theology, might aspire. In the preface to his series of lectures in English published as *The Reconstruction of Religious Thought in Islam*, Iqbal states that 'Classical physics has learned to criticize its own foundations. As a result... the kind of materialism, which it originally necessitated, is rapidly disappearing; and the day is not far off when Religion and Science may discover hitherto unsuspected mutual harmonies'.[15] Iqbal's approach is one that views the opposition between religion and science as resting on an out-dated understanding of both. Even the systems of science issue ultimately from a human, subjective source. It is of course a collective, historically accumulated subjectivity but the same could be argued for the truths of religion, honoured as they are by communities of belief. This historical withdrawal into subjectivity of philosophy and thought generally is something that was first explained systematically by Hegel. Iqbal's recognition of this, derived from his reading of the neo-Hegelians, informs his world-view and in particular his notion of the Ego or Self as the ultimate source of activity in any discipline. In his seventh lecture on Islam, Iqbal divides religious activity into three broad periods: 'Faith', where religion is accepted

---

[14] Ibid, p. xix.

[15] *The Reconstruction of Religious Thought in Islam* (1934; repr. New Delhi: Kitab Bhavan, 1984), p. vi. These lectures were originally delivered at Madras, Hyderabad and Aligarh.

unconditionally; 'Thought', where religion is subject to rational understanding, and 'Discovery' where "metaphysics is displaced by psychology" and "religion becomes a matter of personal assimilation of life and power;[16] and the individual achieves a free personality, not by releasing himself from the fetters of the law, but by discovering the ultimate source of the law within the depths of his own consciousness".[17] This movement towards a rational subjectivity, where the subject acknowledges itself as the ultimate source of the religious and political apparatus looming over it as objective independent entities, has much in common, though is not identical, with Hegel's characterisation of history as progressing towards freedom. Once again there is a noticeable time-lag between the philosophical expression of certain values in the West and their emergence in the East. Hegel was writing on the eve of the violent bourgeois revolutions about to engulf Europe, and many of the bourgeois values he espoused were already in the process of realisation. The movement towards subjectivity and emphasis on individuality that Hegel had discerned was beginning to be expressed in literature, for example in the increasingly popular form of the novel which was by definition a 'subjective' literary form, expressing the world-view of that sacred entity, the 'individual'. When Iqbal proclaims the sanctity of the Ego or Self, this is not an empty Utopian gesture but a powerfully charged affirmation conscious of its own embodiment of the recent history of Western political upheaval from feudalism to bourgeois hegemony. To affirm individuality meant opposing the social structure which had belittled it. But, although the values centred around bourgeois individualism reached Islamic culture in the later nineteenth century, it was mainly the intellectuals, like Iqbal, and later Ráshed, who bore the brunt and shock of this impact. Iqbal's

[16] Interestingly, this is something of a reversal of Comte's periodisation of religious and secular thought: the differences between the two schemes are symptomatic of the underlying differences in secularisation between Europe and the Islamic world.

[17] *The Reconstruction of Religious Thought in Islam*, p. 181.

powerful intellect recognised that Islam's regeneration was not to be sown on the soil already ploughed by Western social advances nor on the barren soil of traditional 'self-annihilation' on which Muslims lay exhausted. It had to come from within, from a Self whose dynamism could institute a new and peculiarly apt synthesis of Eastern and Western elements, an Islam which, while bypassing the sleeping monster of religious fundamentalism, could absorb and integrate new knowledge in the perpetually reinterpreted light of Divine revelation. Iqbal's poetry tends to dramatize these features of his thought, as shown by this ghazal:

> Your song, distraught nightingale, is still unsure
> Let it ripen in your breast a little more
>
> Reason is mature if it follows prudence
> But Love which kneels to prudence is immature
>
> Love leapt into Nimrod's fire, defying fear
> While Reason watches from the roof, still unsure
>
> Love falls to action at the messenger's sign
> While Reason falls back at Understanding's door
>
> Love is freedom, an overhauling of Fate
> But at Fortune's high temple you still implore
>
> The saqi frowns on 'moderate' excuses:
> 'Your heart still inquires what the journey is for'
>
> Who pass round the wine are non-Arab. But mine
> Is Arab wine: yet they shrink from what I pour
>
> From the garden the breeze brings news of Iqbal
> Trapped in a new net, the bird writhes on the floor.

What is striking about Iqbal's ghazal is its use of traditional images which are here imbued with a new sentiment. In Urdu poetry the rose has often been emblematic of the cruel beloved, presented alongside the lamenting nightingale, the disappointed lover. Western readers will recall Oscar Wilde's beautiful manipulation of this bond in his story 'The Nightingale and the Rose'. But here Iqbal is not speaking as a distraught lover. He is lamenting the failure of a broader kind of love, the ishq or basis of the Ego. He is exhorting the Muslims of the world to awaken from their passive inertia. In one sense this first couplet is addressed to himself: he acknowledges that his message, his new insight, is not completely developed. On a deeper level he might be addressing the entire symbolic history of the nightingale in Urdu tradition. Iqbal also uses to new effect the traditional imagery of saqi, wine, goblet and Nimrod. The 'messenger' in the fourth distich is Muhammad and the acceptance of his command means not bypassing Reason but recognising its limits. This is in keeping with much modern Western thought which stresses the limitations of Reason. One may think of Kant's system where pure speculative reason can apprehend only the phenomenal world of the senses and it is the moral faculty or 'practical' reason which apprehends 'noumenal' objects such as God. Other instances of this line of thinking would be Nietzsche's scepticism as to objectivity, existentialism and indeed the more recent trends of literary theory such as deconstruction. The penultimate distich implies that Indian Muslims with their Western educational background shrink from Iqbal's appeal. But this ghazal is as semantically open as the previous one: it is the reader who supplies the specificity.

Under Iqbal's mighty shadow Urdu poetry progressed. But in the 1930s there arose two movements which regarded even Iqbal's poetry, together with his appeals for 'Pan-Islamism', as out-dated and Utopian. The 'Progressive Writers Association', formed in London in 1935, drew up an overtly political manifesto for literature. This manifesto, supporting the Indian independence movement against British rule, was Marxist in its political aspirations and advocated 'scientific rationalism' and realism as opposed to 'revivalist tendencies' in literature.[18] In comparison to Iqbal's language, these formulations are somewhat crude, based as they are on blunt dichotomies of past and present, 'scientific' rationalism and faith, and class. Ironically, though, it was precisely the depth of Iqbal's philosophical vision which led him away from literary 'realism',

[18] Quoted in Mohammed Sadiq, *Twentieth-Century Urdu Literature* (Bombay: Padmaja Publications, 1947), p. 62.

just as Western modernists such as Eliot and Virginia Woolf had reacted against realism in the wake of philosophies which had demonstrated that 'objectivity' was a problematic concept. So, whereas the Western modernists were progressing from realism towards symbolism, the Indian Progressives felt obliged to veer away from symbolism, which had been the essential mode of Urdu poetry (including Iqbal's), and to step into a 'realistic' mode. An alternative way of assessing these advances would be to recall that much Western Marxist theory calls for precisely the same thing: a turning away from endless speculative subtlety in order to focus on material and political realities.

The leading lights of the Progressive movement included the poets Faiz Ahmad Faiz, Majaz, Kaifi Azmi, Josh, Firaq and Ali Sardar Jaffari, as well as the esteemed fiction writers Munshi Prem Chand and Saadat Hasan Manto. The political fervour created by this movement is well exemplified in the poetry of Faiz, where traditional symbols are infused with a political content. Faiz's language, moreover, tends to be more literal than that of his predecessors. The other modern movement in Urdu literature, spearheaded by Ráshed, took this literalisation of the language to its current extreme. It has long been the contention, and in my view a valid one, of Marxist critics that a poem reflects its social and ideological determinants far more profoundly, and sometimes unbeknown to its author, in its form than its overt content. In formal terms, the modernisation of Urdu poetry was not merely a superficial matter of writing in so-called *vers libre* or writing about 'everyday life'. The most profound change in Urdu verse was the relative 'literalisation' and 'desymbolisation' of the language: it is this which underlies the other advances such as the blurring of distinction between poetry and prose, the extension of poetic subject matter to encompass 'individual' experience, a focusing on the present instead of the past as well as an attention to detail and continuity of description. It was essentially under the aegis of an organisation formed in 1939 called *Halqa e Arbab e Zauq* ('Circle of the Friends of Taste') of which Ráshed was a member, that formal experiments made an impact on Urdu poetry as a whole. What divided the ideals of *Halqa* from the Progressives was that its members considered themselves to be writing as 'individuals' rather than as adherents to a political cause such as Communism.

The most brilliant stars in this modernistic firmament were Ráshed and Miraji. Miraji was by all accounts an unusual man. Bohemian in dress and outlook, flaunting always his defiance of convention, he won Ráshed's admiration as 'the most remarkable poet of our time . . .'[19] What Ráshed valued in Miraji was the profundity of his modernistic outlook, embracing psychoanalytic and French symbolist as well as Indian influences.[20] In the poetry of Ráshed, these features are integrated into yet more subtle forms of modernity.

Ráshed's Poetry: Its Nature and Development

Perhaps the most salient feature of Ráshed's poetry as a whole is its progression: like any other Urdu poet, Ráshed began by composing ghazals. But he ended by writing freer verse which was not only remarkable in its subtlety but carried Urdu poetry to unprecedented extremes of innovation. This development was quite conscious as shown by these comments in an unpublished manuscript of Ráshed's:

> ...my poetry...has passed through three...stages. Earlier poems, in *Mawara (Beyond)*, are characterized by youthful exuberance. Most of the poems are...so-called love poems, but underneath exists a consciousness of social and moral repressions surrounding the youth which make it impossible for any love to flower. In the... second period, the latter theme finds more concrete expression in some anger and resentment against political domination, social and moral inhibitions and religious dogmatism. This further led to a broad humanistic ideology...Poems after 'Iran men Ajnabi'

---

[19] Coppola, MS., p. 11.
[20] Coppola, MS. pp. 11/12.

['Foreigner in Iran'] have developed a somewhat global context...They concern..*Man* as such...[21]

A poet, however revolutionary, is obliged to begin with the tradition he or she inherits. It is a mark of Ráshed's stature that he did not indulge in an empty imitation of Western models: this would have produced an artificial poetry which corresponded to no real issues in his social context. Rather, Ráshed integrates both Eastern and Western influences into a self-questioning verse form. This is exemplified by even his earlier poems such as 'The Poet's Past'. Here Ráshed uses the ghazal to question the ghazal form. The poem promises to be a ghazal, beginning with the usual rhyme scheme *aa ba*, but this rhyme scheme breaks down:

These passion-rousing tales of yesterday's
A youthful harp's vexed and wandering lays:
Bold portraits of all my ruined pleasures
Or mirrors of my youth's first wretched phase
A colourful ghazal, praising Leila's hair
And worship of Saleema's magic eyes
Fond writing of Shirin, sentimental airs!
All, a past story of tear-floods and sighs!
Where are you now Leila, and Shirin, where?
Are you, Saleema, still gasping Love's air?[22]

Seemingly simple, the poem is actually a subtle comment on poetic tradition. The ghazal's crumbling form, its breakdown in terms of rhyme, coincides with the increasingly ironic tone which degenerates into sarcasm. However, the sarcasm is directed not at traditional poets but at the legendary beauties such as 'Shirin' and 'Leila' who typically comprise their subject matter. These characters are themselves brought to life as objects of a direct address by the poetic persona, only to hear their own death sentences passed: they are now redundant. They are brought to life in the sense that, instead of being treated as merely the objects of a history of poetic tradition,

[21] Ráshed, *A Brief Note on My Poetry*, unpublished MS., p. [1].
[22] This translation reproduces the rhyme scheme of the Urdu poem.

they are raised to the level of subjects who embody their own lifeless status in that ossified tradition. In a sense, this is poetry addressing its own heritage from within, without the intermediation of any reference to authors. The communication between Ráshed and the legendary characters exists to be overheard, especially by traditional poets. Evident here is the influence of T.S. Eliot's essay 'The Three Voices of Poetry' which Ráshed had translated into Urdu. A degree of self-consciousness is typically built into the traditional ghazal which tends to elaborate on fixed themes. But Ráshed extends this to a historical self-awareness, opening up a dialogue between himself and the past. The traditional elements are there but under the light of a transformed vision. This is a simple example of how modernistic insights can be internally articulated within a tradition rather than being merely imposed from without.

Much Western modernist poetry, symbolist in technique, was an implicit reaction against naive realism. Another token of Ráshed's skilful integration of Eastern and Western influences is that his poetry develops from an initial symbolism inherited from the Urdu tradition, moving through realism to a new, more complex symbolist technique. The lines cited above show the old symbolist mode of Urdu poetry co-existing with but subordinated to a more realistic mode. The earlier poems, mostly in *Mawara (Beyond)*, are written in this realistic vein. Love is viewed as something painful and impossible, as in traditional Urdu verse. But the pain here is a real wound, sustained from an actual relationship. And the impossibility of love is an insight derived after the event: it is not simply the defining form of love. In these early poems love is viewed, in a Donne-like fashion, as a refuge from the world's harsh toil. The lovers comprise a world in themselves. But whereas Donne hyperbolically insists on the reality of such a world, Ráshed acknowledges its artificial and escapist texture. In 'Near the Window' the two lovers are spectators, watching through the window the sordid and empty life around them. This life revolves in a sense around the mosque and the beautiful imagery of the minaret, the highest point of the mosque, suggests that there was once hope here, a hope now devastated:

> I too, like others in the city
> Come out, after each night of love, to
> All this rubbish
> ...Knowing my helplessness, I peer again
> Through this window
> At the minarets of the city's mosque
> When they kiss the red sky a sad farewell

But although love is a refuge, the 'I too' indicates that the lovers are inescapably part of this 'rubbish'. There is another, political, reason for love's impossibility. In Ráshed's eyes, the atmosphere of political oppression and subjugation surrounding him dwarfed the indigenous Asian spirit in psychological terms also: one's thoughts and emotions could not be freely expressed or even felt to their full capacity. What meaning could even the experience of love have in such a repressive, constraining context? In his poem 'Events', Ráshed urges this apprehension on his lover:

> For you, Life's a luscious cluster of dreams
> For me, a wakeful vigil

Typically, this poem is resigned to the actuality of love as a purely physical union, empty of spiritual content: 'If our souls cannot, let these lips meet'. Love, in these early verses, is something surrounded by shadows, and is marked by contrasting imagery of spreading darkness and joyful scenes of Spring. The traditional images of clouds, stars, rose-gardens and dreams inhabit a disillusioned landscape darkened by the skies of religious recrimination. Love can also involve conflicting allegiances. In another poem, the persona is torn, like some of D.H. Lawrence's characters, between love and hate for his lover, who happens to be English. The truly modern element here is that, as with Lawrence, love and hate are not viewed as two coexisting emotions but as aspects of the same complex passion:

For a moment I imagine:
You're no love of mine
But a spinster on some city's shore
And I, a soldier, enemy of your country

In 'Revenge', love itself becomes a weapon of hatred. It takes on the form of an alarming and perverse self-parody:

I remember even now the naked body
Of a foreign woman, from which
My lips took all night
Revenge for the helpless owners of our homeland

No doubt a feeble and pathetic gesture: to 'conquer' a woman whose race has conquered one's homeland. But in its very mean-mindedness the poem shows the profound psychological impact of foreign domination which has disturbed the foundations of the most ordinary human impulses, turning them into something brutal and ugly. The poetic technique here is a realism which attempts to express immediate experience. But, as with Eliot, the notion of 'experience' itself is viewed as complex and problematic. Eventually, this perception of its complexity will deepen so as to infuse the form as well as the content of Rásheds verse; it will lead to a questioning and rejection of realism as an adequate mode of conveying 'experience'.

Perhaps the most interesting example of Ráshed's treatment of love, as well as of his developing technique, is the remarkable four-part poem 'Hassan the Potter'. Again, to appreciate this poem's innovative nature, one must consider Ráshed's starting point as bequeathed by Urdu tradition. We could profitably read this poem as a modernised version of an Arabic and Persian poetic form known as the *masnavi*. Traditionally this was a long romantic narrative poem in rhyming couplets. Like other verse forms, it had its peculiar conventions, beginning with lines in praise of God and the prophet Muhammad, moving to a eulogy of love and a recounting of the circumstances of the poet's composition. Other conventions might include a description of the poet's first encounter with the beloved, a rebuke by Fortune, description of the woman's state, the meeting routine of the lovers and the grief of their separation.

Ráshed's poem echoes the traditional masnavi as the ghost of its own living voice; it inhabits a vacancy, hinting at the masnavi as its vestigial structure. The poem is written not in couplets but in free verse (which in Urdu retains a certain regularity of metre). But there are more significant departures from the traditional form. In the first place, there is no straightforward narrative. The poem begins with the persona directly addressing his lover, Jahan-zad:

> Jahan-zad, below in the street, before your door
> Here I stand, heart aflame, Hassan the potter!
> This morning in the bazaar, at old man
> Yousuf the perfumist's shop, I saw you
> And in your eyes was that fire
> In whose longing I have wandered mad for nine years
> Jahan-zad, for nine years I have wandered mad!

Instead of a chronological narrative, the starting point is a retrospective present. Here is a curious mixture of archaic and novel elements. The repetitions 'I have wandered mad', the reliance on word sounds for semantic effect and the image of the distraught lover enraptured by the beloved's eyes are all echoes of traditional emphases. Yet the touches of realism, such as the perfumist's shop and the bazaar or market-place, leave the beloved untouched. There is no description of Jahan-zad (as there would be in a masnavi) and of course she is not really being addressed except as a rhetorical occasion for the poet's self-reflection. The poet is effectively addressing himself and Jahan-zad is reduced to a symbol within this internal dialogue. Her symbolic content, however, is not pre-given by tradition but is created in the course of the poem.

It is no accident that the lover Hassan is a potter. Adding to the complexity of love's treatment here is a wealth of symbolism behind the 'potter' image. Nearly all the invocations of this image in the Old and New Testaments are used analogically to illustrate God's authority. The potter's wheel is the cycle of existence, and God, the potter, has the power to make and unmake Man, his earthen vessel:

> Nay but, O man, who art thou that repliest against God? Shall the thing formed say to him that formed it, Why hast thou made me thus? Hath not the potter power over the clay...?[23]

Persian poetry, such as that of the mystics Rumi and Hafez, also makes use of the 'potter' image: it is, among other things, a reminder of Man's ultimate origin and destiny, dust. But these tend to be passing references. Perhaps the most sustained emphasis on the analogy of the potter occurs in a text long familiar to Western readers, the *Rubaiyat* (quatrains) of Omar Khayyam. The final section of the *Rubaiyat*, as translated by Fitzgerald, is entitled the *Kuza-Nama (Book of the Pots)*. It is here, in the descriptions of the pots, rather than his boasts about wine, that Khayyam's antinomian and unorthodox tendencies are brought to light:

> Listen again. One Evening at the Close
> Of Ramazan, ere the better Moon arose,
> In that old Potter's Shop I stood alone
> With the clay Population round in Rows.
>
> And, strange to tell, among the earthen Lot
> Some could articulate, while others not:
> And suddenly one more impatient cried—
> 'Who is the Potter, pray, and who the Pot?'[24]

Ironically, Khayyam questions Divine transcendence enlisting the very image traditionally used to affirm it. But this is strictly a questioning of transcendence (with the implication that the boundaries between God and Man are not so rigidly drawn), not a denial of God's existence. Khayyam's rejection of the outward forms of religion certainly brought him into

[23] Romans, 9, 20–21. See also Isaiah, 29,16; Jeremiah, 18,6; Psalms, 2,9; Lamentations, 4,2 and Daniel 2,41.

[24] Edward Fitzgerald, *Rubaiyat of Omar Khayyam* (London: Penguin, 1989), p. 58.

conflict with religious orthodoxies but does not amount to atheism. Ráshed is but one culmination of this unorthodox trait in 'Islamic' literature. There are some similarities between Khayyam's images, of the persona standing alone in an old man's shop at a particular time of day, and the beginning of Ráshed's poem. Ráshed of course has gone several steps further on the path of heterodoxy, the relation between potter and pots being secularised: the pot has become the potter. The potter now is not God, not even a questionable God, but a living human being who has the same problematic relation to his pots as God previously had with humans.

So, in the same way that Eliot's *The Waste Land* refuses to yield a 'literal' meaning, depending as it does on traditional voices speaking through its own voice, Ráshed's poem derives its meaning not simply from what the words 'say' but from their status as self-conscious participants of a history of symbolism. As well as the 'potter' symbol, the images of 'wine' and 'moth' are used to new effect, with the new meaning containing or gesturing toward, the traditional one:

> How shall I forget
> That night's wayward delight?
> Was it wine or the trembling of my hand
> Caused a glass to topple and break?

Wine is still associated with love and the breaking of the glass could signify the failure of love but these symbolic repercussions are constrained within a more factual, naturalistic context. Again, the 'moth', the traditional distracted lover drawn to the candle, becomes the occasion of a new metaphor:

> Time
> Is the moth, crawling
> On walls, mirrors, cups, glasses
> On my jars, pitchers and pans
> Like crawling time, perhaps
> Hassan the potter will return, his soul burning

In this reflection, which is both generalised and particular, one can see how layers of symbolism interact in a complex way.

Each image, potter, wine and moth, has been literalised so as to undermine its privileged symbolic status. However, not only does the symbolic ghost lurk behind its literal replacement but this combination of literal voice and symbolic echo itself yields a new symbolic framework based on the interaction of the new meanings. Hence the moth, formerly the lover, is time crawling towards Death which seduces it. The mark of Death, in the form of time, is etched on the previous emblems of love and mystic revelation (the wine glasses), as well as on the dust and clay which composes the pots, the created beings. The movement of time has a human embodiment in the emotions ('soul burning') of the creator, the potter himself. This new framework of symbolism is no longer communal but esoteric and highly individualised. It is the outward aesthetic form of the poem's iconoclastic affirmation of Self which underlines the poem's more obvious themes of love, art and livelihood:

> Before all else I am 'Self'
> If only we exist, still
> I am
> Self before all things

That this assertion of individuality should be so overt, rather than simply reflected by the form, betrays the urgency and desperation of a poet living in a society where little emphasis was placed on the individual.

Iran men Ajnabi (Foreigner in Iran) was mostly written during Ráshed's stay in wartime Iran. Not surprisingly, they are more overtly political in the sense that they make numerous allusions to contemporary political events, such as the invasion of Iran by Britain and Russia. But Ráshed's invective is not reserved for foreign powers only. He also strikes out at the Shah of Iran:

> King Reza
> A foreigner's salutation to you!
> An Indian soldier's salute!
> How can you see me?

Your eyes are fixed along the horizon of Elburz
Here, below your statue
I stand on the gleaming stairway

The statue, of course, betokens the Shah's self-aggrandisement, one of the themes of this poem. This piece is part of a series of cantos, wide-ranging in subject matter and style, including elevated passages with Persianised diction as well as colloquial dialogues revealing the sordid intimacies of private lives. These cantos are also structured around a contrast between the glories of ancient Persia and her relative stagnation in the modern world:

Theatre of a field of flowers
Is Iran no more?
...Iran is now a wailing old woman
With beauty long withered
At the desolation of Ctesiphon Persia sheds tears
Naushirwan and Zardasht and Darius
Farhad and Shirin, Keikhusrau and Kaikbad
We are a story, they its characters
We a caravan, they its leaders
Whose tombs lie below the earth

Iran is a history unfulfilled, a narrative begun and looking in vain for completion. A striking poem of this period is 'Deserted Sheba', which plays on the multiple symbolism of Solomon both in the Qur'an and the Bible:

Solomon, head in his hands:
From where now will come a joyful envoy?
From where, which jar, will come wine into
The bowl of old age?

The man of wisdom, now perhaps signifying the United Nations[25], is pictured as a beggar left to rule a desolate kingdom.

The title of Ráshed's third collection, *La = Insan (X = Man)*, indicates that he was moving away from immediate political

[25] Sheila Ráshed tells me that Ráshed once offered this particular interpretation.

dilemmas towards a more general perception of the difficulty of defining Man. If Man's nature is not prescribed by religion, and society offers an incomplete explanation, there is something mysterious and problematic about existence. A Western reader might see this as merely passing from one mystifying image to another: yet it has something in common with existentialist views of Man and the many Western philosophical visions which deny Man any 'nature' as such. By this stage of Ráshed's writing we can see that what were previously perceptions on the level of content have become so deeply thought through that they are sublimated into the formal structure of his verse. Ráshed is now concentrating as much on the inadequacy of language to convey experience as the most accomplished of Western modernists. A good example is a poem entitled 'That Lonesome Word (Which Desires to Unite with Meaning)':

> Our limbs, raised in prayer toward the sky:
> (Don't look at the sky!)
> Is this to bypass harm in a fragile place
> Or a pretext for hiding from our deprivations?
> That ancient and high god (Paradise in truth)
> Shall one day deliver us from god
> For on this earth we are as a lonesome Word
> (But it won't be a world like this) silent and speaking
> Which lives in desire for accord with meaning
>
> Grant us the dream of M'arri
> (Grant each a smile fitting his vision's breadth)
> Grant us the vexation of M'arri's soul...

The poem is a dialogue between its bracketed and unbracketed parts. It is not a conversation between two people but a debate within one mind whose stream of thought is continually interrupted. The speaker is unidentifiable, talking in a vacuum, with no perceptible audience except his own unconscious. The 'lonesome Word' was *the* Word, God, and is now Mankind. The only place where it can unite with meaning is Man's mind: the gap between word and meaning embodies Man's doubt, the subject matter of the poem. God undergoes an ontological shift, from being a reality to subsisting as a concept. This challenges the traditional irreducibility to language of the Logos, which is not simply a linguistic designation of God but betokens the rational order of Creation itself. In other words, the self-referential nature of the Logos is denied, and the Word is reduced to merely another word in the system of language. This is an implicit recognition that the 'reality' of God has been passed down through language. Even here, it is stripped of referentiality, to signify Man's blindness to even the phenomenal appearance of God. But, paradoxically, it is a blindness towards which the persona aspires: the ancient Arab philosopher al-M'arri, also a religious sceptic, was blind and Ráshed paradigmatically invokes him to exemplify the continuity of religious illusion. The entire poem, in fact, paradoxically takes the form of a prayer, a prayer for blindness rather than insight, the end being M'arri's profound spiritual vexation as opposed to a shallow certainty. These are themes which are continued in a yet more experimental vein in Ráshed's fourth book, significantly entitled *Guman ka Mumkin (The Possibility of Doubt)*. Here, the breakdown of language and narrative extends even to individual words and Ráshed's versifying is sustained by neologisms.

What is at issue in Ráshed's individualism, and the nature of his modernism in general, is not a question of 'freedom', whether religious or political—India, after all, is as much a democracy as any other nation claiming this title—but the complex phenomenon of secularization.[26] This phenomenon, the decline of religion in terms of both institutional power and thought, is one that occurred almost exclusively in the West. It began effectively with the Protestant Reformation in Europe; this contributed to the displacement of a Roman Catholic emphasis on 'other-worldliness' (whereby this life was regarded as transitory and essentially worthless) by a supposedly more rational 'this-worldliness', whereby the

[26] One of the few excellent accounts of secularization in the Islamic world can be found in Bryan S. Turner, *Weber and Islam* (London: Routledge, 1974).

present world rose in significance. Many Enlightenment and nineteenth century developments of thought extrapolated these implications: science, Biblical scholarship, evolutionary and political theories all contributed in some ways to loosen the grip of religion on the human imagination. But all of these were epiphenomenal of economic developments: it was essentially the rise to political power of the bourgeoisie on the waves of industrialisation, urbanisation, economic rationalism and individualism, that threw back the defining shores of religion. Islamic cultures simply did not undergo these economic changes until much later, hence the correlative displacement of religious thought, in terms of both intellectual institutions and personal psychology, was also delayed. In as much as secularization has occurred in Muslim nations, it has been mimetic of Western secularization. It has been a surface phenomenon, with no economic foundation until the relatively recent past. It should be said that Islam was never as exclusively 'other-worldly' as Roman Catholicism. It has always had a practical and rational emphasis. But equally, it has never been divested of its 'other-worldly' strain which is, as with most religions, its *ultimate* foundation. So, when we encounter secularised attitudes in Eastern poets such as Ráshed, it is worth remembering that such figures were part of an intellectual élite nurtured on exposure to international currents of thought. Ráshed achieved some externality to his literary tradition but had the acuity and knowledge to criticise that tradition from within. Even Ráshed's unorthodoxies are a testament to the way Islam defines the lineaments of a person's psyche in Muslim culture: it is always the starting point and one would rather blame God for Man's misfortunes than seek a wholly secular explanation. Praise and blame may be opposed but they exist at the same spiritual depth, a depth which Ráshed has reclaimed for Urdu letters.

BEYOND

ماورا

Near the Window[1]

Lamp of Love's chamber, awake!
Wake from this joyful floor of soft dreams
Your body still tired from night's pleasure
Come by me, lover, near the window
And see with what passion dawn's rays
Kiss the minarets of our city's mosque
Whose height brings to mind my Age-long desire

With your silver-white hands, my lover
Open those wine-dark,[2] bewildering[3] eyes
See this minaret
Watered by early light
Beneath its shadow, I remember
A mournful, penniless priest
Drowsing in a dark, hidden corner
Like a useless god
A devil,[4] distressed![5]
Here is[6] the stain of three hundred years
An indignity without cure.
See the crowd in the market place
Moving, an endless flow

---

[1] In an unpublished manuscript entitled 'Notes on Twelve Poems by N.M. Ráshed,' Ráshed states that this poem 'portrays all round helplessness and frustration of a people living in a decadent society and under a foreign rule, when leaders, particularly religious leaders, are caught in an inertia, the intelligentsia suffer from an impotence, and the common people, living in slummy tenements, have contented themselves with a marginal existence of routine and drudgery'. p. [1]. Hereafter cited as *Notes*, MS.

[2] Literally, wine-coloured.

[3] Lit. madness-causing.

[4] عفریت literally refers to a giant, ogre or demon.

[5] Lit. sad.

[6] 'Here is' is inserted.

As *jinns*[7] in the wastelands
Emerge at early evening, bearing torches
A bride-like figure sits
In the corner of each man's heart
The tiny lantern of Self flickers
Without strength to burst into
Spinning flame
Among these are the poor, the sick
Below the heavens tyranny[8] marches on.[9]
I an old, weary, ambling horse
Ridden by Hunger, hard and robust
I too, like others in the city
Come out, after each night of love, to
All this rubbish
The sky is turning where
At night I return to this same house
Knowing[10] my helplessness, I peer again
Through this window
At the minarets of our city's mosque
When they kiss the red sky a sad farewell.

---

[7] Spirits, said in the *Qur'an* to be born of fire (lv, 14–15). They are often thought of as evil and as assuming fantastic forms.

[8] Or 'cruelty'.

[9] Lit. goes on.

[10] Lit. seeing.

In the Dead Silence of a Shoreless Night[1]

Sometimes on your bed, my love
In the dead silence of a shoreless[2] night
Limbs go lifeless with passion
And beneath this burden of pleasure
The mind sinks[3]
Into a quagmire of some wilderness
And somewhere nearby
Sleep, like a bird fearing
Predators at Winter's dawn
Prepares its wing, and screams
In the dead silence of a shoreless night

Sometimes on your bed, my love
Longings crawl in your bosom[4]
Like a long-oppressed Negro
For a moment I imagine
You're no love of mine
But a spinster on some city's shore
And I, soldier, enemy of your country
Soured without sleep
Long fugitive from my unit
Seizing this hope of endless pleasure
This is the thought that holds me
Sometimes on your bed, my love
In the dead silence of a shoreless night.

---

[1] Ráshed's note: 'A study in ambivalence to show how love and hate, and pleasure and pain go together and often, imperceptibly, merge into each other. Mark the political innuendo of the imagery'. *Notes*, MS., p. [1].

[2] Lit. boundless, immense.

[3] Lit. becomes.

[4] Lit. in the mountain of your breast: this, of course, sounds somewhat overstated in English.

Revenge

I remember her face, her features
I remember a room
A naked body beside a fireplace
A carpet on the floor, on the carpet a bed
An image of bronze and stone
Laughing, in the wall's niche
In the fireplace sparks crackle their fury
At the images, so inert[1]
And a reflection stains the white, white[2] walls:
The souvenir of Western rulers
Whose swords placed here
The foundation stone of Europe

I remember her face, her features
I remember even now the naked body
Of a foreign woman, from which
My lips took all night
Revenge for the helpless owners of our homeland
I remember.

---

[1] Lit. In the fireplace, the crackle of sparks/Furious at the lifelessness of those images. بےحس can also mean 'unconscious' or 'senseless'.

[2] Can also mean 'luminous', 'radiant' or 'splendid'.

City[1] *of dreams*

(Sonnet)[2]

Lover, there's the other shore![3] Let me go
    I'll never return to this desolate
Place. You cannot come, let me greet my fate[4]
    Alone I will go and like an arrow
For God's sake, be selfless and let me go
    Alone I'll suffer whatever hardships wait
But there alone I'll find some path, some gate
    To peace.[5] I've no strength to argue, let me go
From the city of dreams what sounds beckon?
    Who comes to greet me from the other side?
What's his message, only God can reckon
    If I stayed here, my soul would not abide
Friend and lover, release me, let me go
    From these empty shores: you cannot follow.

---

1 Also 'town' or any peopled place.

2 Ráshed uses a slight modification of the Petrarchan sonnet form which I have attempted to reproduce here.

3 پار can also mean 'opposite bank' or 'other side'.

4 Lit. Don't insist on coming, just let me go. The original order of the second and third lines is inverted here.

5 Lit. But if I go to that side, perhaps I'll find peace.

Poet's Past

These passion-rousing tales of yesterday's [1]
A youthful harp's vexed and wandering lays
Fearless portraits of my ruined pleasures
Or mirrors of my youth's first wretched phase
A colourful ghazal praising Leila's[2] hair
And worship of Saleema's magic eyes
Fond writing of Shirin,[3] sentimental airs!
All a past story of tear-floods[4] and sighs!
Where are you now, Leila, and Shirin, where?
Are you, Saleema, still gasping Love's air?[5]
Death's calm spreads over my past
Which maelstrom,[6] my lanterns,[7] now quenches you?
My verses, visions of my 'lost paradise'
Even now, your light falls in the heart's landscape
To lay you down on beauty and love, I seized you
From the company of moon and stars.

---

[1] Lit. last night.

[2] This alludes to the romantic epic *Leila and Majnun* by the twelfth century Persian poet Nizami.

[3] This refers to another romantic epic by Nizami, *Khusrau and Shirin*, the central theme of which is Farhad's love for Shirin. Farhad was a gifted engineer who had been promised Shirin if he could cut through Mount Bisitun. At Khusrau's command, he was given false news of his beloved's death and was left to lament his loss and wasted effort.

[4] Lit. tears. The image 'tear-floods' derives of course from John Donne's 'A Valediction: Forbidding Mourning' which also mocks a conventional publicised sentimentality of love.

[5] Lit. still tired on the path of love.

[6] Lit. storm of oppression.

[7] The lanterns refer to the conventional images and styles of love poetry.

Hassan the Potter[1]

(A long poem in four parts)

I

Jahan-zad, below in the street, before your door
Here I stand, heart aflame,[2] Hassan the potter!
This morning in the bazaar,[3] at old man
Yousuf the perfumist's shop, I saw you
And in your eyes was that fire
In whose longing I have wandered mad for nine years
Jahan-zad, for nine years I have wandered mad!
Lost in that desire[4]
I never turned toward my sad pots
Those images of my restless hands
Lifeless creations of dust and colour and oil
Now they whisper:
'Where is Hassan the potter?
Creating us,[5] he's become a god!'
Jahan-zad, nine years passed over me
As time treads over some buried city
Dust in flower-pots
Whose aroma I'd fondly breathe

---

[1] The first section of this poem is actually contained in *X = Man* and the remaining three in *The Possibility of Doubt*. For reasons of continuity, however, it seems better to keep together the entire poem and to present it at this point in the overall selection of Ráshed's work; the poem displays a mixture of traditional and modernistic strategies. Ráshed's note states that 'Love is no doubt important to an artist; but his art and his means of livelihood, are even more important. Even in its great intensity, love is meaningless in the end unless it turns into an *élan vital* for the artist and leads him to still better creativity'. *Notes*, MS., p. [2].

[2] Lit. head-burnt

[3] The marketplace.

[4] Inserted.

[5] Lit. he, through us, through his own act.

Was now laden with stones
Goblet, enamel, cup, pitcher, lantern, flower-vase
All hope of an art to express
My worthless existence
Lay dead,
And I, Hassan the potter, dust on my head
Dishevelled hair, prostrate on the potter's wheel
Like some downcast god
Creating pots in a dream world of being and nothing[6]
Saw, in your bright eyes of Caucasus[7]
That fire
Through which my body and soul became wayfarers
Of cloud and moonlight
Jahan-zad, that dreamy night in Baghdad
The bank of the river Tigris
The ship, the closed eyes of that sailor
For some weary, disheartened[8] potter
One night alone was alive[9]
Which even now claims his
Spirit, his body
Only one night's joy the river's wave has granted
In which Hassan the potter sunk, never to emerge

And now, Jahan-zad, each day
That unlucky one comes to haunt me
Prostrated on the potter's wheel
It shakes me by the shoulders
(That wheel which year to year[10] was my only hope of livelihood)
'Hassan the potter, come to your senses

---

[6] Lit. in a flow of dreams of unreal being and nothing.
[7] The meaning of افق is obscure here.
[8] Lit. grief-filled.
[9] Lit. کہربا = yellow, amber, electricity.
[10] Or, for years, for a long time.

Cast an eye on your ruined[11] home
How shall the bellies of these children be filled?
Hassan, Love's fool![12]
Leave that sport of the rich
And look to your own house.'
In my ear that sorry voice fell
Like one calling down a whirlpool to a drowning man
Yes, that lake[13] of tears fed life to flowers[14]
But I, Hassan the potter, was enchanted
By ruins of the city of illusions
With no sound, no motion
No shadow of a bird in flight
No trace of life

Jahan-zad, in your street today
Against night's chilling darkness
I stand restless[15] before your door
Through the window, those enchanting eyes
Peer at me again
The age, Jahan-zad, is that potter's wheel on which
Like enamel, cup, pitcher
Lantern and flower-vase
Humans are created and destroyed
I am human, yet
These nine years have passed in the shape of grief
Hassan the potter is today a heap of dust
Without sign of moisture.
Jahan-zad, this morning in the bazaar at
Yousuf the perfumist's shop, your eyes
Have spoken once again

---

[11] Lit. empty.
[12] Lit. love-stricken.
[13] Lit. reservoir.
[14] Lit. 'was a reservoir of flowers'.
[15] Lit. head and hair restless.

Breathing moisture[16] into dust
Perhaps dust will waken into clay.[17]

Who knows the expanse of desire, Jahan-zad, but
If you wish, I'll become once more
That same potter whose pots
Were the pride of every palace and quarter, every city and village
Adorning the dwellings of rich and poor[18]
Who knows the expanse of desire, Jahan-zad, but
If you wish, I'll turn once more towards my sad pots
Those dried pans of being and nothing
Towards hope of an art to mirror my livelihood
From that being and nothing, from that colour and oil
To strike again the sparks by which
The ruins of hearts are illumined.

[16] Lit. light or lively moisture.
[17] Lit. Perhaps this would turn the dust to clay.
[18] Lit. With which the dwellings of rich and poor were made bright.

II

Jahan-zad,
How shall I forget
That night's wayward delight?
Was it wine or the trembling of my hand
Caused a glass to topple and break
You were hardly[1] surprised!
Your windows had suffered cracks before
Jahan-zad,
Returning from Baghdad to
These pots, these earthen pans
I consider[2] how you've been a mirror before me
In public, at the window, at the head of an ermine bed
Where nothing has surfaced
But my own face
The fear of my own pathetic solitude
As I write to you
That mirror and its one face
Lie in my hand
And why should this mirror, my tablet, not be
Etched with tears?[3]

Jahan-zad,
Will I ever feel again
The delights of that wanton night?[4]
Time
Is the moth, crawling
On walls, mirrors, cups, glasses

---

[1] Absent in the original but used here to recreate the mood of the line.

[2] Lit. 'I think, I think' repeated.

[3] Omitted here are some repetitions which would sound overdone in English.

[4] Lit. Will you ever bring to me again?

On my jars, pitchers and pans
Like crawling time, perhaps
Hassan the potter will return, his soul burning.

Returning, I wonder
If I'm not the web of the spider deprivation
Being woven on this cottage roof
This dark cottage where I muse, prostrated:
My poor ancestors
Left but a single trace
This is the story of their craft, their livelihood.
Now I've returned, that luckless fortune
Comes to stare at me
And will not cease[5]
In this my cottage, there's nothing
A game of simple love
We play
In the growing hollowness of night and day
Crying, singing, laughing
Together
Keeping alive and running the heart's blood, nothing more.
Words have boundaries, Jahan-zad, meaning a boundary
Love, youth, tears and smiles[6]
All have boundaries
(Have then the pain of loss
And isolation no boundary?)
In my cottage
How many aromas
Hover about me
Like the smell of that one night they cling
What dusty aromas linger here
Aromas of my poverty, loneliness

---

[5] Lit. continues to stare for a long time.

[6] Lit. the flow of smiles.

Memories and desires
But still in this cottage, dirty, dishevelled
There's nothing
Though sometimes the song of birds drifts from far trees
And fragrance from gardens of olives and figs
Then I'm cheered
And glad to have bathed and emerged[7]
Else, in this house, there's no bed, no perfume
Not even a fan
The love you know so well
Is beyond my strength

You'll laugh, Jahan-zad, at my strangeness
My bounty with emotion
My worship of things
My seeking a wealth which is not mine
You who laughed that night at my vacillation
Laugh, laugh again at my torn-ness![8]

Yet who from love has found anything beyond self?
Jahan-zad, all questions
Of love have but one answer:
A lover[9]
It's enough that the heart's voice should echo
Jahan-zad
It was the voice in a corner of my heart
Echoing
At the shores of my art,[10] frozen in time
Your eyes were an ocean whose endless gaze

---

[7] Lit. saying, behold, I've bathed and come out.
[8] Lit. duality, two-sidedness.
[9] Lit. each question of love has no answer except a lover.
[10] Lit. At the shores of the centuries of my frozen art.

46

Froze those shores in their stilled centuries[11]
Ocean, mirror of my Self
Of the faces of my pots destroyed and created
Of each art and its worshipper
The mirror.

---

[11] The original lines are modified considerably here.

III

Jahan-zad
The pool of a caravanserai[1] at Aleppo, that silence of night
Where we swam,[2] embracing
Waves outspreading,[3] like circle around circle
We swam the whole night, holding
Each other body and soul
Swimming with delightful fear
Just as water swims in tears
Content in each other, against the tide[4] of age
You teased: 'Your longing, Hassan, has
Dragged you even here!'[5]
But a fear now swims in my heart:
I left my body in that pool at Aleppo
Yet no dualism haunts me[6]
Even now I've faith that body and soul are one
Faith which absorbed
Me into Self
Before all else I am 'Self'
If only we exist, still
I am
Self before all things
How could I betray this?[7]

---

[1] A Persian name for an inn with a courtyard where caravans of travellers spent the night.

[2] Lit. kept swimming.

[3] It seems best to regard محیط as a verb; a nominal translation would make unintelligible the relation between lines 2 and 4. Its subject here is 'that silence'.

[4] Lit. against the decline of age. 'Tide' better accommodates the antagonistic connotations of خلاف .

[5] Two lines are omitted here, as superfluous.

[6] Lit. I have not the fear of dualism.

[7] Lit. If I exist, how can I betray my Self?

Women like you
Complex beyond
Unravelling
To say I'd unveiled your depth[8]
Would be self-deceit
Woman's nature[9] is self-satire
For which we are no answer

(Who is Labib? Words of him all night
Stole from your lips
Who was it kept pulling at your hair
Tearing at your lips
As I never could
Yes,[10] whether myself or Labib[11]
If a rival, why, for the pristine delight of your self-knowledge
Which walks, like the morning, in several voices?[12]

---

[8] Lit. To say I have unravelled you.

[9] Or 'fabric, construction.'

[10] Lit. this is true.

[11] There are several possible interpretations of this passage. In the first place, 'Labib' is a name. Secondly, 'Labib' is an adjective of لُب which means 'wisdom' or 'essence'. Thirdly, it could be Rashed's own derivation from لَب (lip), whose Sufistic connotations are cited by A.J. Arberry, based on Muhsin Faid Kashani's *Risala-yi Mishwaq*, as 'the life-giving property of God, and his keeping Man in existence', *Sufism* (1950; repr. London: George Allen & Unwin, 1963), p. 114.

It seems oversimplistic to strip the word 'Labib' of its variety of possible symbolic meanings to the denotation of the mere presence of a third person as a rival. On the other hand, whatever symbolism 'Labib' imbibes through varying traditions, it is evidently personified in this and subsequent passages of the poem. Therefore it seems best to translate it as a name which is symbolic. If رقیب means 'rival', the rivalry is for Jahan-zad's 'self-knowledge' and the entire passage must be read in the light of traditions of rivalry in Urdu poetry. But رقیب can also mean 'one who keeps guard over', as an attribute of God, in which case those connotations of 'Labib' deriving from لُب may be semantically admissible.

[12] Lit. a hundred voices and one voice.

Labib, the negation of every favouring voice!)
Yet ours is not a union of water and dust, nor ever was
These elements have lived always outside Man
Nor was jar or pitcher always born of their union
They can yield but one illusion
Let it be

Jahan-zad, you
He and I
We angles of an ancient triangle
Have always wandered, like the
Turning of my potter's wheel
But found no trace of ourselves
If you wish, I'll break the triangle. But
The spell of the wheel on me is this old triangle
The eyes of my wheel which stare at me
Turning
On pitcher and jar your body, colour, your tenderness
Fell
Your beauty's alchemy
Washing me in a flood of internal light
Citizens of my inner world poured out into the streets
As if hearing the cry of the morning *azan*[13]
All the pots in becoming[14] became 'you'
This meeting's ecstasy has devoured me
This is the crisis of cup, goblet and pitcher:
Cleaved from the essence of water and dust
They achieve the triumph of new direction

(I, a poor potter
What do I know of this extreme

---

[13] Azan means 'call to prayer'. It is normally recited by the muezzin from the minaret of a mosque.

[14] Or, 'in being made'. The original effect is perhaps better achieved by repeating the verb 'become'.

Mystic intuition of
Each cup, goblet and pitcher?)

Jahan-zad
Today I am waiting still, but why
Just as I was for nine wretched years
But now my waiting is not for the river of tears
Nor the wayward night
(I've talked so much of that night of sin
That too is now a sin)
Neither for the pool of the inn at Aleppo
Nor for Death, nor my broken[15] self
Yet in timeless waiting I am bound[16]
As moments arrive and pause in timeless time
So this burden of time has fallen from my shoulders[17]
All the dead, all past faces
The sluggish caravans of all incidents
Have awakened inside me:
The stirring of a world regained
As heaven awakes in the unconsciousness of god
I woke lying on the sand of drowsiness
On the sand lay those pots
—Outside my being—
Shattered forever
In the chasm between me and myself
They became whole once more (like some voice of Providence)
They became once more a timeless dance
A vision of eternity.

---

[15] Or 'defeated'.
[16] Lit. But of timeless waiting a string is bound.
[17] Lit. from my head.

IV

Jahan-zad, how after a thousand years
Scraps of my jars, enamels and flower-pots
Are found
In every alley of a buried city
They are[1] the memory of this ruined city
(By the name of Hassan, a young potter, in a new city
Still making pots, still loving
Was strung by us in the threads of his past
In us (as though he were us) he is merged
You and I were the drops of rain falling
All night (a night stretching a thousand years)
On a window-pane, becoming drops
Of snake venom[2]
Here, before the morning of time
We and this youthful potter
Are strung once more in a dream)

Jahan-zad, how
This crowd, worshippers of the past
Has entered the corpses of these pots
See!
These are the people whose eyes
Never reached into jar or pitcher
Today they upheave once again
The lifeless creations of colour and oil
Will they ever find, below these, the sparks of grief
Which devoured history?
Will they ever hear[3] the storms, the tempests

---

[1] Lit. As though they were.

[2] Lit سانپ لہریں refers to the coiled marks which a snake leaves upon the ground. But here, 'snake venom' seems more compatible with 'drops'.

[3] Lit. find.

Which devoured every scream?
What do they know of the rainbow which brought my colours?
(Mine and this young potter's?)
What do they know of that butterfly's wings
That beauty, the character and features whereby
I shaped the faces of these pots?
See these people, each in his prison[4]
The age, Jahan-zad, is an enchanted tower
With these people in its prisons
The young potter laughs!
These naive savages, their garments torn
By their own stature, reaching for
Some glory beyond reach
What do they know of the demon, inevitable, in my heart's cave
Who forewarned me (and this young potter):
'Hassan the potter, awake!
The pains of prophethood have their day of reckoning
Which approaches your parched cups and pitchers!'
This was the call behind which, Hassan
The young potter
Moves from age to age
Autumn to Autumn
Ceaseless

Jahan-zad, I, Hassan the potter, have
Suffered this pain of prophethood in wilderness
After wilderness
Will these people, a thousand years hence
Reassembling the pieces, ever know
How the colour and oil of my dust and earth
Merged with your fragile limbs
To become eternity's voice?
Through my pores, every pore

---

[4] Or 'each his own prisoner'.

I would absorb
Your expansive embraces
I would make offering
In the temple, the eyes of each future generation[5]
Should they trace the civilisation of these fragments,
so be it
How shall they ever bring back Hassan the potter?
Or count the drops of his sweat
Or find even the shadow of this art's splendour?
Which has grown from age to age
Autumn to Autumn
Which, in the new self of each potter
Ever grows
The shadow of that art through which
None is anything but Love
None is anything but a potter
We are full of knowledge, and know nothing
We are, like god, wholly the gods of our art
(Hopes are shallow and deep)
Faces swim in the eyes of senselessness
Faces never seen
Where could their trace be found?
Who has ever honoured the tradition of grief?

The corpses of these pots
Etceteras of some mortal story
Are our azan, the mark of our inquiry
In the silence of their hour of death they speak:
'We are the eyes which open inwardly
Which gaze at you, seeking out every pain
Knowing the secret of each beauty

---

[5] Lit. those to come.

54

We are the longing of that night's empty room
Where one face, leaning
Over another, like a tree branch
Planted in each man's heart
A rose petal
We are that night's stolen kiss'.

FOREIGNER IN IRAN

ایران میں اجنبی

Distance

Death will come for[1] me, I will die
Death will come for you
That first moonlit night will be a night
Of two half moons
Like two ends of an ancient harp's broken strings
Like shores of two horizons
They vibrate from afar and will not come near
Of mysterious worlds they sing no more[2]
And He who struck[3] the singer from the
Rolling cycles of space and time
Was granted release through eternal sleep

Then why
Could we not hear the march of future ages
To halt their invasion with our breasts
For are not we your destination, your refuge?

Perhaps he thought
He himself would create this distance
(Now a yawning ocean)
That we, exulting
Might find again the forgotten life.
Wretched excuse, self-deceit!

---

[1] Lit. to me.
[2] Lit. Nor does talk of mysteries pass their lips.
[3] Lit. ejected.

But now the distance itself is whispering:
You cannot return to your station
You cannot return
Death will come for me, I will die
Death will come for you
But first this devil shall be
Wiped out!

Desolate Breweries

In the desolate breweries of Marree[1]
I search for that candle[2] whose
Journey[3] through glass, goblet and the Saqi's[4] hand
Has brought the fire[5] of Self to
Corners of minds darkened in grief

Standing before the candle's flame
I peer, through a mystic's eyes
At Marree's desolate breweries
I shiver in the ecstasy of success
Travellers of the world have never known:[6]
Whence came the silver morning of the Nile's Fate
Spreading colour and song
Over the dry centuries of Egypt's sand dunes?[7]
Today I too stand at the edge
Of a Nile's broad banks

Ruins stand
Like the morning of Creation
Believing if the beginning was infinity's soul[8]
Infinity became in time the beginning
Of an apostle's first journey

---

[1] A town in Pakistan, to the north of Rawalpindi.

[2] Lit. wick.

[3] Lit. passing the destinations of.

[4] In Urdu literary tradition, the beautiful woman who pours wine at a poetic gathering, who is also the object of the poets' infatuation.

[5] Lit. lantern.

[6] Lit. have never brought an answer to this.

[7] Lit. The silver morning of the Nile's Fate/Which filled for centuries/ Egypt's dried sand dunes/With colour and song.

[8] Lit. substance.

Ruins, staring like the morning of Creation
Are not sorry to see dead
Their hearts' light.[9]
They know also the rebellion of this candle.

---

[9] Lit. the light of their bosoms.

Nimrod's Divinity[1]

This was the soil[2] of holy men
Where philosophy saw in its early morning sleep
The vision[3] of a fresh breeze, lush fields, and a cool[4] brook
A traveller, reaching here, wonders:
'Was that dream a nightmare?
A nightmare?'

O seeming philosopher
Where are those heavenly visions?
Where the divinity of Nimrod?
With whose broken threads you've been weaving
Your obscure metaphysical net
Today, we despair
Of that faith,[5] action
And love

Who, after all, will ask:[6]
What priceless justice was borne out
By Tartar ravagings?

---

[1] There are passages in the *Qur'an* which speak of Abraham's trials as a youth: he defied the idols worshipped by his father and his people. After smashing these images he was condemned by his people to be burned but he was unharmed by the fire (XXI, 51–71). Although the *Qur'an* does not specify Nimrod, this incident is traditionally associated with this king. It has been conjectured that his capital was in Assyria, near Nineveh.

The Old Testament also speaks of a Nimrod, though existing before the time of Abraham. Begotten of Cush of the generations of Noah, 'he began to be a mighty one in the earth... He was a mighty hunter before the Lord...' Genesis 10, 8–9.

[2] Or 'land'.

[3] Or 'image'.

[4] Lit. heart-cooling.

[5] Or 'certainty'.

[6] Lit. Who, after all, will ask of whom.

Persia, enchanted land of colour, wisdom[7] and song
Arabia, honeyed land of milk, wine and date-palms
They merely sang of their ruined homes
We are the helpless prey of days even worse

Who, in the end, will ask:
Were these houses of iron, timber, stone and silver
Beneath some spell of exquisite harmony?
Why mourn the shattered spell?
What false faith in harmony!
Pitchers and wine-cups are destined to break
As a lover's cheek will lose its colour
But here, ruined hearts
(A ruined Ctesiphon[8] of human quest nobler
And higher than galaxies)
Herald a lament: dead is
Harmony between word
And meaning.

---

[7] Or 'thought'.

[8] The ancient city of Ctesiphon, also known as Seleucia, was the capital of the Parthian empire. The Parthian dynasty in Persia extended from 250 BC to AD 226.

Deserted Sheba[1]

Solomon,[2] head in his hands,[3] and Sheba desolate
Sheba desolate, the home of ghosts
Sheba an abysmal lake[4] of woes
World devoid of grass, greenery and flower

Winds thirsty for rain[5]
Birds of the desert, beaks tucked beneath their wing
And Man, choked on dust

---

[1] Ráshed's note on this reads: 'Solomon in this poem is a conqueror and lover who destroys what he conquers and kills what he loves, and in his old age is left with nothing except a deep remorse and a deep sense of futility. The poem has an allegorical reference to the politicians or dictators of today, who in their keenness to set things in order, often end up with doing exactly the opposite'. *Notes*, MS., p. [1].

[2] Solomon, son of David and King of Israel from c. 970 BC, has quite a complex symbolism both in the *Qur'an* and the Old Testament. The latter repeatedly refers to Solomon's wealth and wisdom, as in this extract: 'So King Solomon exceeded all the Kings of the earth for riches and for wisdom'. I Kings 10, 23. But later Solomon is said to do 'evil in the sight of the Lord...' I Kings 11, 6. He was visited by the Queen of Sheba (called Bilqis in Arabic tradition) I Kings 10.

In the *Qur'an* Solomon is said to have been given knowledge by God (XXVII, 15) as well as power over the wind (XXXIV, 12). Sheba, a city in Yemen, is described in the *Qur'an* as a beautiful gardened terrain, sustained by a dam which made it a prosperous and civilised area until its people turned away from God; the dam was destroyed and this flourishing 'Garden of Arabia' was turned into a waste (XXXIV, 15–16). This seems to be the starting point of Ráshed's poem.

[3] Lit. head on his knees.

[4] Lit. reservoir of woes.

[5] As well as reiterating the traditional life-giving associations of 'rain', the *Qur'an* ascribes various symbolic meanings to the winds: they are heralds of 'glad tidings', purifying agencies (XXX, 46–48) and renewers of life (VII, 57).

Solomon, head in his hands, bitterly dishevelled hair
World-dominion, world-administration, merely
the bounding of a deer[6]

Love a leaping flame,[7] lust the odour of odourless flowers
Speak less of the age's mysteries!
Sheba is wasted for still on her soil
Are footprints of a ravaging conqueror
Sheba is no more, nor her beautiful queen

Solomon, head in his hands:
From where now will come a joyful envoy?
From where, which jar, will come wine into
The bowl of old age?

---

[6] This imagery is reminiscent of the Song of Solomon in the Old Testament: 'The voice of my beloved! Behold, he cometh leaping upon the mountains, skipping upon the hills. My beloved is like a roe or a young hart...' 2, 8–9.

[7] This echoes an image in a love sonnet by Mallarmé (whom Ráshed had read): Le chevelure, vol d'une flamme... [Her hair was once like a leaping flame...].

Which Mystery Do We Unravel?[1]

Lips a wilderness, kisses lifeless
Which mystery do we unravel?
We ourselves become the fuel[2]
For these workshops of the body
Midnight in a dream-stained city, we are shadows
Wandering like thieves through night
Since evening we've been free
Slaves of desire, drinking
Glass after glass, thinking
Some hidden drop might
Signal at last the first secret

Intention easy, words meaningless
Contrived curves of smiles
All the notes of the text
Outlining[3]
Visions of unripe pleasures
Between our bodies not a hair's space
Between our hearts a falling[4] chasm
With eyes and ears close, which mystery did we unravel?
Which mystery do we unravel
When at night we steal from our gloomy caves?

---

[1] Ráshed's note: 'The theme of this poem is a search for futile love or a futile search for love, in big modern cities. Love attained through a long and painful process of "love-techniques", often turns out to be hollow inside'. *Notes*, MS., p. [2].

[2] Lit. firewood.

[3] Lit. from which are made.

[4] Lit. deep.

Is Life seeking new straits
Does the swift god of senility stand over[5] us
Or could it be an ego's ambition?[6]
Which mystery do we unravel?

---

[5] Lit. before.
[6] Lit. expanse of the ego's body.

Foreigner[1] *in Iran*

[A long poem in thirteen cantos][2]

I

Manna and Quail[3]

'Most high God
This, the land of old Joshua[4]
The equitable courts of Naushirwan the Just[5]
The galleries of Sufism, wisdom and literature
Why have these become again a boiling ulcer, implanted
By dark-skinned foes
Why?'

Where is our guilt in this, beloved Persia?
The first Englishman who
Brought and planted the commodities of his trade
On the shores of Hindustan[6]
It's his sin
That our flower-covered country is
Trampled beneath our own dark feet

This city is not our homeland
But the Englishman's plundering
Has forced our liaison with it

---

[1] Or 'stranger'.

[2] A selection of these cantos is presented here.

[3] Manna and quail is the 'heavenly' food which sustained the Israelites in their wandering over the wilderness.

[4] The Old Testament successor to Moses.

[5] The most illustrious member of the Sassanian dynasty, who acceded to the throne of Persia in AD 531. He was called Kisra by the Arabs and Chosroes by Western commentators and is remembered for his wisdom, breadth of mind and benevolence.

[6] This is the Urdu term, deriving from Arabic, for India.

We are a lizard
In the height of his civilisation, the plunderer
Imagining:
'Asia is a widow, barren and rich
Ever pressed for her vast unguarded wealth
Her pining bosom awaits me
And those Asians are a tired[7] generation
Moving swiftly towards their death'
These Indians, hungry and bare-footed, have
Mystic knowledge but
No sense of where custom or tradition will lead[8]
After despoiling houses
After enduring a thousand hardships
After spilling our blood
They[9] think (if they think at all)
Perhaps only their arms
Can buy salvation for the East
From the foreigner's depthless[10] cruelty
Imagining
The chance which rolled them into
Your homeland
Might set ablaze the vicious fuel[11] of war[12]
To spread throughout the world

My many comrades
In the outstretched, dark intervals
Of their careworn, useless lives

---

[7] Lit. slothful.
[8] Lit. no knowledge of the destination of custom and tradition.
[9] This refers to the Europeans.
[10] Lit. boundless.
[11] Lit. germs. The original would make awkard reading in English.
[12] Lit. the arena of those germs which ever give rise to the vicious smell of war.

On the pathways
Seeking some luscious calf
Sometimes, from behind windows
Beyond all sense, they
Jump on helpess fluttering moths
The hand of the beggar
Outspreads before them
Hoping their bounty might
Quicken a piece of bread into essence of
Manna and quail:
This tender hand
May point the way
To the road where
Craving knows no delight

Seeing these sights you think
These cowardly tyrants are
Fettered in the unjust 'love' of Englishmen
Whose fraud putrefies[13] this city
There is no unjust love
Only a chain
A single iron noose
Spread
From one shore of the East to the other
From my homeland to your homeland
A single spider's web in which
We entangled Asians wrestle for freedom
From the blood-spilling morning of the Mughals[14]
To Europe's life-draining sunset.

---

[13] Lit. is like a boiling ulcer.

[14] 'Mughal' is the Persian word for Mongol. Deriving initially from central Asia, and extending over northern China, Siberia and Persia, the Mughal Empire was extended to India in 1526 by the Emperor Babar.

We wrestle for recovery
In a single cureless pain
Cleaving our souls
This alliance has cost us
One another.[15]

[15] Lit. This costly union has barred us till now from reaching out to one another.

IV

Alchemist

King Reza[1]
A foreigner's salutation to you!
An Indian soldier's[2] salute!
How can you see me?
Your eyes are fixed along the horizon of Elburz[3]
Here, below your statue
I stand on the gleaming stairway

It's said that
In devotion[4] and high reverence
You had this statue[5]
Carved in your reign
By Europe's famed sculptors
Erected in this square, a monument
To your faith in your nation's applause![6]

---

[1] Reza Khan, proclaimed Reza Shah Pahlavi in 1926. He strove to limit the Shi'ite clergy's influence in political and cultural matters. He furthered female emancipation, introduced new laws regarding marriage and divorce, improved education and was staunchly nationalist. But his reign was stamped with cruelty as well as greed for personal possessions (to which Ráshed alludes later in this canto). He abdicated in September 1941 and died in exile in 1944. He was succeeded by his son Muhammad Reza.

[2] From 1943 to 1947 Ráshed held the rank of Inter-Services Public Relations Officer in the Indian Army, serving in Baghdad, Tehran, Basra, Cairo and Colombo.

[3] A range of mountains in the north of Iran.

[4] Lit. Through extreme devotion, granted you by your venerable nature.

[5] Ráshed is referring to a statue in Tehran.

[6] Lit. in your nation's recognition of [your] kindness.

King Reza
Heir of the house of Darius and Cyrus[7]
Bent on beautifying this realm
You could hardly leave its care to Providence[8]
After the Shahriwar[9] disasters
Here is the realm you've left behind:[10]
A lifeless corpse
Clawed by three vultures[11]
With inspired[12] lust!

Who does not know
What searing love of country and heir
Fired your sense and thought and act!
You were the builder who
Had no business with foundations
Night and day with infinite speed, you were set
On turning your eyes from the cracks

---

[7] Cyrus I and Darius III both belonged to the Achaemenid dynasty. Cyrus proclaimed himself King of Persia in 546 BC; Darius III acceded to the throne in 336 BC.

[8] Lit. to God.

[9] The sixth month of the Persian calendar.

[10] Lit. Today it [your realm] lies in your footsteps.

[11] Ráshed may be thinking of Iran's relations with Britain, Germany and Russia during the earlier part of World War II. At the outbreak of the war, Iran had declared her neutrality. But after Germany's declaration of war on Russia in June 1941, all three of these powers were aware of the strategic importance of Iran. Germany had conducted an internal campaign in Iran to prepare a receptive atmosphere for her armies once they penetrated Iran. This German activity ultimately provoked an invasion of Iran by both Russia and Britain in 1941. The first American troops arrived in Iran in December 1942. The impact on Iran of the war was harsh, causing scarcity of food and most types of goods as well as high inflation.

Ráshed may also have had in mind the post-war situation: in 1944, British, American and Russian oil companies began making overtures for concessions and participation in the Iranian oil enterprise.

[12] Lit. new and growing.

On ornamentation of ceiling, wall and door[13]
And who could forget your
Foreboding speech at Nadir's[14] tomb:
'In Nadir was every excellence
Except discretion: men foresaw
His designs'

True, had Nadir not
Made known at night his terrible
Plans for morning, his murder
Might have passed him by[15]
But your secrecy surpassed all limits
Besieging you
In your own thought
A prison through which your eyes wandered
Finding everywhere your own face only
Glimpsing each belief
Through a blind window (you called it inspiration)[16]
Blocking
Any ray of thought[17] for your nation

The result? One day
You had much to say (as usual)

---

[13] This is perhaps Ráshed's comment on the rapidity of some of Reza Shah's reforms, as well as his taste for splendour.

[14] Nadir Shah who, from lowly origins, rose to take the Persian throne in 1736. Nadir has the reputation of being the last great conqueror in Asia: he led successful campaigns against Bokhara, Khiva, the Turks and captured Delhi, the Mughal capital of India. His remaining years, however, proclaimed him a tyrant in the wake of his increasing cruelty and greed.

[15] Nadir's later injustices led to a widespread rebellion, which included the Kurds of Kuchan. In 1747 Nadir marched on Kuchan and was killed in his camp by some of his own tribesmen. Having heard that they were to be seized and killed, they entered his tent late at night and assassinated him.

[16] Or 'revelation'.

[17] Lit. Which housed not even a single small inlet.

And no listeners anywhere, all deaf![18]
It's two years since you left this earth[19]
You have no history in their hearts[20]
Your image lives (if at all) in these words:
'An alchemist, who
Promised to bring forth gold[21]
But stole
Even the people's copper and silver'.

This Tehran, in your dreams
Was a fairer copy[22] of Paris
But in its streets there flows
At evening a flood of face-sellers
Even now in dance-halls
Sparkle the lamps of parties
Here, tired through dancing or drunk with wine
Lovers of the homeland
Sing in unison
To music of dulcimer, tambourine, guitar and flute
Songs of 'The homeland, the homeland'
But who knows whose land this is?

Today, even Paris has taken its form, wretched![23]
On it has fallen a spark[24]
The flames of the reed-bed devour it endlessly!

---

18 Lit. And if there were, they became deaf.

19 Ráshed was evidently writing in 1946.

20 Lit. Your memory is wiped from their hearts.

21 In the original, direct speech is used here but this reads awkwardly in translation.

22 Lit. revised version. Iran had close connections with France at this time.

23 Or 'helpless'.

24 لبرقِ خرمن literally means 'spark to a pile', implying that the spark ignited the pile.

VII

Hand of the Tyrant

In this caravanserai at the bridge
Each meeting of ours
Became a deeper loneliness[1]
But today's separation is
A wound in the soldier's heart
Not so deep as to rob his sleep
But it stings enough to forbid
Any hint of a smile across his lips

Farewell, O rose of Poland
Congratulations: today you journey to a new world!
Wherever your companion[2] is taking you today
Poland, for certain, will not be there
But in that land, the brilliant star
Beaming in the eyes
Of freedom-loving masses
You'll find some release
From unreal hopes
From this toil, night and day, for foreigners
There, the harmony of colour and language
Will console
And, away from this grief, from the terrified caravans
Of down-trodden refugees
There will shimmer the mirage of destination
You'll always be a wasted[3] bough
But on that new soil
Hope will rise, for green leaves and lush flowers

---

[1] Lit. Every meeting remained in the form of a stronger loneliness.
[2] Lit. equal, peer.
[3] Or 'broken', 'snapped', 'dilapidated'.

I know well how
You've been trampled, driven out, homeless
The grief you voice
Is not yours alone
It grows even now
From speck to colossus

You were no Nazi
Had no alliance with the Fascist mind
Your crime was to seek peace
And satisfaction in the goals
Of your beloved soil
Of the moonlit night, bright party, wine-glass and jar
You sought
A deep sleep at dawn
On the easy, warm cushion of a lover

And your punishment (O Lord):
Before your helpless eyes, your lover's corpse
And, in a foreign prison forced
To labour in Russia's snowy regions
For bread-scraps left from the night
And now, this year-long
Service in military camps
This abject beggary, this
Broken, aimless life
Whose past, present and future
Prostrate themselves on desolate ground[4]

Beyond the world of reality
Is a world of thoughts, of dreams
Which becomes finally the way of destiny
But this extreme of tyranny

---

[4]Lit. Whose past was desolate/Now present and future also lack meaning.

Settled endlessly
On your thoughts, on your dreams
Like moss of despair
How can I forget, O nightingale of Poland
Those songs, rustic songs of Poland
Which, in the idle nights of the soldiers' camps
I have heard on your lips
Where, for the sake of a dark-eyed Indian
A Polish woman's heart beats, reaching for
The Indian's musk-scented[5] beauties
Songs, which inspire this feeling:
The night-wandering spirit of the West
Surrounded by uncertainty, chased
With soft footsteps
Hovers now over[6]
The ravaged nests of our
Last Eastern refuge of dreams

Here is a symbol
Of that night-wandering spirit:
A dispersed refugee caravan, condemned
By the tyrant's hand to
Wander
Through expanses of the West and East
Farewell, moonlight of Poland
For us too, one hope remains:
In this foreign land
Are these meagre resources
In the wind's path, one blade of grass
This will be uprooted soon, this caravan also
Of the wretched and destitute
And the dawn of peace shall once more
Become its footprint.

---

[5] Or 'musk-coloured', 'jet-black'.

[6] Lit. Now, at last, hovers constantly over.

VIII

Dervish[1]

This Winter evening
Below, in the avenue
Before my windows
In the frozen darkness
Comes the sound of ice-bound feet:[2]
'O Lord
Will tonight also your stony eyelids be lifeless as rock?'
The avenue stretches far, an inferno of deep darkness
And I
Plunge down
To uncover the meaning of sounds

This dervish
Whose ancestors

---

[1] This is a Persian word meaning 'poor' and is used primarily for Sufi ascetics and mystics. Sufism began about AD 800 and Sufi ascetic practices, including the wearing of garments made of coarse wool (suf), were intended to facilitate a mystic union with God. By the thirteenth century, there had appeared regular international orders of Sufism. The disciples of the thirteenth century Persian poet and Sufi, Maulana Rúm, formed the Maulawi or Mevlevi sect: these were the 'whirling dervishes' who performed a sedate mystical dance which recalled the order of the heavenly spheres.

Ráshed's note on this poem reads: 'Deals with the problem of beggary in emerging Asian countries. A comparison is drawn between the beggar of today, who is left helpless on the streets, with his predecessor, the Sufi, who had, so to say, chosen the path of beggary, because it gave him the superiority of indifference to the world at large. For the latter there was some consolation in the mysticism of Rumi and Hafiz, but the beggar of today is only face to face with laws which forbid him from begging, without providing an alternative. The modern statesmen who make such laws are hardly aware that they can turn against them too, possibly in the form of a people's revolution'. *Notes*, MS., p. [1].

[2] In the original this is posed as a rhetorical question.

Tired and dying
On the sand of a past desert
Were just like him
Rolling empty-handed in dark earth
Achieving indifference through
Acceptance
Borne ever aloft on ascetic wings[3]
Eager for survival of Royal pomp
In whose shadow poverty might seem glorious[4]
Walking among the dead, prompting them:
'Sleep on!
The dawn of tomorrow does not exist!'
Their mystic freedom could but urge
Imperial power to
Admit its limits[5]
But who knows the proper limits?

But today's beggar
Deprived from eternity
Ruined, like his ancestors
By the politics of time's intriguers[6]
Faces a boundless obstacle[7]
Even the rusty visions
Of Maulana Rúm
Of the divine meditators of Shiraz[8]
Can't help

---

[3] Lit. They thought lifelong deprivations to be their wings.

[4] Lit. For the sake of the glory of poverty. It is not impossible that Ráshed intends a reference here to the brilliant Safavid dynasty of Persia which was founded by a family of Sufi Sheikhs in the sixteenth century.

[5] Lit. Should not exceed its limits.

[6] Or 'chess-players'.

[7] Lit. His real hardship is this.

[8] A city in Iran.

'Lord
Will tonight also your eyelids be lifeless as stone?'
Why must you still say this, you
The age's downtrodden

How it pleases you
That a thousand tongues have become
Your tongue[9]
To engulf, like flames from every side
The dreamy palaces of Kings

Politics thought it would
Silence you
How should it know that
Where lips close
Arms and hands will open?

Why is the sound of those frozen, heavy feet
Suddenly hushed?
O scourges of the new constitution
Of the novice East
The beggar's calm
Will not quell poverty's cry.

---

[9] Lit. That a thousand tongues speak on your behalf and, having become your tongue.

X

Oil Merchants[1]

For one black mole the towns of Buqara and Samarqand![2]
But where now are Buqara and Samarqand?
Buqara and Samarqand are lost in dreams
Hidden[3] in the veils of an azure silence
Their doors closed to travellers

---

[1] Ráshed's note states that this poem 'records impressions of the Russian manoeuvres during the war to coerce Iran into signing an agreement with them for the exploitation of her oil, criticises the expansionist policy of big nations, under the garb of economic co-operation. Mark the Arabian Nightish imagery of the poem, and a comparison with the conditions which led to India's subjugation by the British'. *Notes*, MS., p. [1].

[2] In one of his ghazals Hafez, the great fourteenth-century poet of the Persian city of Shiraz, had addressed his mistress in these words:

> For a mole on the cheek of my darling,
> Which the breezes of Shiraz have fanned,
> I would gladly surrender Bukhara,
> Or give back to its Khan Samarkand.

The great Mongol conqueror Timur (known in the West as Tamerlane) captured Shiraz in 1388. According to legend, Timur desired to see the poet who had such apparent disdain for Buqara and Samarqand which were the conqueror's two seats of power. Hafez came before Timur shabbily dressed. Timur exclaimed that he had conquered half the world and yet the wretched poet standing before him was prepared to sacrifice his principal cities for a beauty spot on his mistress's cheek. Hafez is said to have won the favour of the conqueror with his reply: 'It is my excessive generosity which has reduced me to my present state'. Translation of Hafez's verse from J.C.E. Bowen, tr., *Poems from the Persian*, (Blackwell, 1948), p. 84.

Buqara and Samarqand were towns in Transoxiana (south west of modern U.S.S.R.). During the Samanid dynasty in the latter ninth and tenth centuries, they were brilliant centres of Islamic culture, housing such men as Avicenna, as well as being host to the beginning and flourishing of Persian literature as we know it. It is evidently a decline from this status of grandeur to which Rashed is alluding.

[3] Or 'drunk'.

Like the eyelids of a beautiful sleeping woman[4]
Preserved from the lash of Russian 'Pantheism'
Two beauties!
Forget Buqara and Samarqand
And think now of your shining cities
Of the roofs, doors and terraces of
Tehran and Masshad[5]
Look to the
Pleasant fountains of your age of sense and action
And these beautiful metaphors of your new hopes
Make high[6] the low walls
Of these splendid cities
Post your sentries at every tower and rampart
And in your homes, silence all sound[7]
Save the wind
For outside, beneath the city walls
Robbers have long been pitching their tents
Dressed as oil-merchants
Tomorrow or tonight at dark
They'll come as guests
To your houses
To drink from the goblets
Of the banquet night
They'll dance, sing
And laugh impulsively, noisily
To warm the blood of the gathering!

---

[4] Lit. Like a sleeping woman with a moonlike brow. This is an ornamented metaphor for a beautiful woman.

[5] One of Iran's most populated towns. It has long been the capital of the province of Khorasan. Its name means 'burial place of a martyr' and it is noted for housing the holy shrine of 'Ali Reza, eighth Imam of the Shi'a line, who was poisoned in the ninth century.

[6] Lit. strong.

[7] Lit. silence the lanterns of all sounds.

But when dawn breaks
Then you will dig
Graves for your dead with your eyelashes
Then you'll shed tears
On the ashes[8] of merriment's[9] stage
We also have shed tears
—Though the black mole is worthless
That deep oozing ulcer which arose on earth's cheek
From Europeans' murdering lust—
We also have shed tears
Cities, like liquid shadows, dissolving in waste[10]
Falling roofs and doors
Minarets and domes
But time is an arch
And our enemy passes through its curved flanks
Rolling down its lower horizon
Imprisoning and whipping
Our naked, lean bodies
Our tyrant begins to
Sweat in his own fire

Place your hand in mine!
Place your hand in mine!
For I have seen
Rays on the peaks of Alwand[11] and Himaliya[12]
Through them at last will break
A Sun
Longing for this, Buqara and Samarqand
Have long been beggars.

---

[8] Lit. On the shore of the burnt ashes
[9] Or 'entertainment', 'feast', 'banquet'.
[10] Lit. Wasted/dissolving cities, like liquid shadows.
[11] A mountain on the western side of Persia.
[12] The Urdu name for the Himalaya mountain range in North India.

XIII

Theatre[1] of a Field of Flowers

Theatre of a field of flowers
My eyes were focused on the 'theatre'
My ears fixed to rise and fall of 'music'
But still, my heart was counting
Woes of Arabia and Persia
Theatre of a field of flowers!

Theatre of a field of flowers
Is Iran no more?
This masterpiece of love—'The doomsday of Iran'!
Iran is now a wailing old woman
With beauty long withered
At the desolation of Ctesiphon Persia sheds tears
Naushirwan and Zardasht[2] and Darius
Farhad and Shirin,[3] Keikhusrau and Keikbad[4]
We are a story, they its characters
We a caravan, they its leaders
Whose tombs lie below the earth
Theatre of a field of flowers

---

[1] Also 'sight', 'scene', 'spectacle' or 'stage'.

[2] Also known as Zarathustra, of which 'Zoroaster' is a Latin corruption. This figure was the founder of the Zoroastrian religion in ancient Persia. Authorities dispute the period of his existence, placing it as far apart as the fifth and tenth centuries B.C.

[3] See p. 38, fn.3.

[4] Keikbad was the first monarch of the Keianian dynasty, which belongs to a period of Persian history which is still overlaid with legend. Keikhusrau was a later monarch whose champion, according to some legends, was Rustam, the subject of poems by the Persian Firdausi and Matthew Arnold.

Theatre of a field of flowers
Till when shall dirges be our tongue?[5]
Harsh the path[6] from sorrow to everlasting sorrow
They were all large-hearted and wise, worshippers
Of a benevolent God
They knew and followed virtue's hard path
Today we're far from gratitude
Those great emperors
Their ancient rank and dignity, their lost pride
Were the fabric of our defeat
Self-annihilation was their fate
And we the mourners of their fate
Mourning still in their longing
Theatre of a field of flowers

Theatre of a field of flowers
Bride of the youthful future, covered in veils
Eager eye, sad from hasty men
Our new dreams are
Not the nightmares of the past
But dreams of a new Man, of
A world of endeavour, not Ctesiphon
Not the palaces of Faghfoor and Khusrau
These are not the asylum of the new Man
New settlements and new rulers
Theatre of a field of flowers.

---

[5] Lit. Till when the irritation/displeasure of dirges?
[6] Lit. destination.

X = MAN
لا = انسان

The Marriage of Abu Lahab[1]

Abu Lahab's wedding: what a night!
His bride came bearing
Firewood on her head
A garland of snakes about her neck[2]
No care had she for make-up or colour
Oil or perfume![3]

The crowd's wrath flared out like
A tyrant's lashes on naked bodies
Dead was the young boys' applause
And the dancing dainty feet of coy girls
Stilled was all music and festive song

Feeling the crowd's heat, Abu Lahab
Gripped the reins, spurred his horse
And Vanished[4]

---

[1] Abu Lahab was an uncle of the prophet Muhammad and belonged to his tribe, the Quraish, in Mecca where Islam was born. He was nicknamed 'father of flame' after his fiery temper and ruddy complexion. One of the most relentless opponents of the incipient religion of Islam, it was his practice to throw insults and curses at the prophet. An early Meccan Sura of the *Qur'an* entitled *Lahab* explains how this 'father of flame' will himself be consumed in flames (CXI, 1–3). Abu Lahab died shortly after the defeat of the Meccans by the prophet's followers at the battle of Badr in A.D. 624.

[2] Abu Lahab's wife was equally spiteful to the prophet; she would tie up thorns with ropes of palm-leaf fibre and strew them where she expected Muhammad to walk. The *Qur'anic* passages state that when Abu Lahab is consumed by flames, his own wife shall carry the firewood and will wear a twisted rope of palm-leaf fibre about her neck (CXI, 4–5).

[3] A repetition of three lines is omitted here.

[4] Lit. There was no news of him.

No news of him: years turned
To dust, scattered forever[5]

Amassing rubies, pearls in foreign lands
Abu Lahab returned like a ghost
Countless[6] curious eyes peered from
Old windows, a dense throng
Of young and old rushed
Out to see
Abu Lahab's procession

'Abu Lahab!' someone called, a figure[7]
From that night's splendour, a puff
Or whirlwind from the sands of fancy
A ghost of ruined love
'Abu Lahab! Are you he whose bride
Came with firewood on her head
And snakes about her neck?'

Abu Lahab shrank back
Spurred his horse towards
Oblivion.

---

[5] Lit. A span of many years, having become a cloud of dust, was scattered.
[6] Lit. a thousand.
[7] The word 'figure' is inserted to make the sense clearer, implying that it is Abu Lahab's bride from that night who is now the 'ghost' addressing him.

The Death of Israphael[1]

Mourn for the death of Israphael
Favoured[2] of the gods, lord of speech
Undying soul of human voice
Boundless call[3] of the heavens[4]
Now still as silence:
An unfinished word

Come, mourn this untimely sleep of Israphael
Who rests by his clarion as if
Thrown[5] by storms onto the shore
On the sand, in the glittering sun
He sleeps by his trumpet
His turban, hair and beard
Grimy with dust
Whose layers were once being and nothing

---

[1] According to *Qur'anic* tradition Israphael (known as Raphael in Judaeo-Christian tradition) is the archangel who will blow the trumpet on the Day of Judgement. At the first blast, everyone in heaven and earth will swoon and lose consciousness; at the second blast the revived earth will shine with God's grandeur, *Qur'an*, XXXIX, 67–70.

Ráshed's note to this poem reads: 'Israfil, in Islamic mythology, is the angel, who will blow the trumpet on Doomsday. Here in this poem he has been used as a symbol of all sound and speech. The death of Israfil would therefore be the end of all speech, all expression, and thus of all culture. The dictators who try to suppress human expression today, are only creating an ultimate vacuum'.

[2] Or 'esteemed'.

[3] Or 'sound'.

[4] Perhaps an allusion to the 'deafening noise' which the *Qur'an* states will introduce the Day of Judgement (LXXX, 33).

[5] Lit. spat out.

His trumpet, once time's breath,[6] fast and slow,[7]
Lies far from his lips
Lost in screams and agony[8]

Mourn for the death of Israphael
An Incarnate tumult, an incarnate song
A sign
Of unseen voices stretched from time's beginning
To eternity

For Israphael's death
Angels mourn, stood circle within circle
Man is lean, with dusty hair, and
Divine eyes dark with grief
No singing streams down from heaven
No trumpet-blast from the divine world

Israphael's death
Has starved the world's voices[9]
Of singers, of harps
Stilled are the strings of listening hearts
What shape can the dancer's dance now take?
Hushed are the floor, doors and walls of society[10]

---

[6] This could refer to the alternatingly fast and slow sounds flowing from the trumpet; it could also be alluding to the *Qur'an*'s affirmation that after death our perception of time will change; the time spent in this world will seem brief (XXIII, 112–14).

[7] Lit. delay and swiftness.

[8] Or 'complaints'.

[9] The *Qur'an* states that after the trumpet-blast, all sounds will be humbled: there will be a stillness of awe and reverence (XX, 108).

[10] Lit. gathering.

What shall the city's orator[11] cry?[12]
Silent the domes and minarets of mosques
What net shall the hunter of thought now cast?
Speechless the birds of flight[13] and mountain

Israphael's death
Has killed
The listener's ear, the speaker's lip
The seeing eye and knowing[14] heart
His very breath[15] shaped the cries[16] of dervishes
And all dialogue between saints
Who now sit recluse and choked
Vanished now is 'tanana hu',[17] lost is 'ya rabha'[18]
Fled[19] is every voice from lane and alley
And lost our final refuge

The world's time seems glazed,[20] in sleep
And all voices swallowed
In oblivions whose silence deafens out
All memory of ideal beauty[21]
Memory of one's name

---

[11] Or 'preacher'.
[12] Lit. explain/tell.
[13] Lit. destination.
[14] Lit. wise, shrewd.
[15] Or 'existence'.
[16] Lit. the 'ha' and 'hu' of dervishes. These exclamations were used by Sufis at meditation or during a *Qawwali*.
[17] This is the name of a tune invented by the poet Amir Khusrau, born at Patiali in India in A.D. 1253. He composed in both Hindi and Persian. Traditionally, it was sung just before the end of a Qawwali.
[18] 'Ya rab' means 'O God'. The phrase is exclamatory. 'Ya rabha' is the plural form.
[19] Lit. lost.
[20] Also 'petrified'.
[21] Lit. That one cannot remember perfect beauty.

Israphael's death will
Sweeten the speech-suppressing dream[22]
Of our world's dictators
Dreaming of a sovereign world[23]
Where oppressed whispers[24]
Might still be heard.

---

22 Lit. The world's dictators will continue to watch their dream of suppressing speech.

23 Lit. a world of sovereignty.

24 Lit. whispers of the oppressed.

Mirror Void of Sense and Knowledge[1]

Mirror, void of sense[2] and knowledge
How should we turn its nothing[3] to being?
Being rests[4] on the frenzied search of night and day
How should we show a mirror's heart
A mirror?

We mourn that heart's futile expanse
Dry of foliage[5]
Where the lately blossomed flower gives no smell
Men, though beautified with eye, lip and ear
Are deprived
Of excitement,[6] of the light
Of you and I
Here wine cannot flow like tears
Nor intoxication glow
Nor lovely frenzy stir the mirror's pure heart[7]
Imagination[8] finds no path through its emptiness[9]

---

[1] Ráshed's note: 'The mirror reflects everything, the whole life for that matter, but nothing grows in its vastness, nothing is dynamic inside it. The mirror in this poem may be taken for a symbol for a man who lives under a borrowed light or of a nation which lives on foreign alms'. *Notes*, MS., p. [2].

An interesting comparison could be made between Ráshed's poem and Sylvia Plath's poem 'Mirror' which also affirms a mirror's passive and lifelessly objective registering of sense-impressions.

[2] i.e. sensation.

[3] Lit. non-being.

[4] Or 'is dependent on'.

[5] Lit. deprived of verdure.

[6] Lit. the pleasure of excitement/tumult.

[7] Lit. Nor the beauty of frenzy in the mirror's pure heart. It can also mean 'tumult', 'commotion' or 'agitation'.

[8] Lit. thought.

[9] Lit. Nor in its empty heart is the path of imagination.

Mirror, void of sense and knowledge
How shall its nothingness assume being?[10]
The mirror is a sea, silenced
At Creation[11] by a magic hand[12]
Image on image enters, hoping to
Break[13] this lonely heart's charm;[14] to
Shatter its death-like calm

In its dark[15] world the mirror
Hears the drops of time's dew
Watching reflections, its speech entombed[16]
Like a buried city.

---

10 Lit. How shall we turn its non-being into being? I have varied slightly Ráshed's exact repetition.

11 Lit. at the beginning of time.

12 Lit. a magician's hand.

13 The words 'through its essence' are omitted here.

14 Or 'magic'.

15 Lit. mysterious or full of secrets.

16 Or 'tongue-tied'.

Death, meet these
The simple-hearted, whom
Neither prayer nor wine can please[1]
Accomplished in no art or science
Whose hearts have no room[2]
For books or machines
Whose eyes can face[3]
Neither earth nor space
Unbelieving men!
Death, don't hide[4]
Greet them!

And you! You too come forward
To meet Death!
Come, *nouveau-riche* beggars
and don't hide your begging bowls
You've no commerce left with Life
Laugh with Death and make it laugh!
Slaves of money, of time, come!
Death, these scum[5]
Are negative men, more
Negative, less men:
Cast a gracious glance on them!

---

[1] Lit. Accomplished in neither prayer nor wine.

[2] This is absent in the original but is used to recreate the effect of terseness and finality which depends in part on the rhymes.

[3] Inserted.

[4] Lit. Don't veil yourself from them (or don't be bashful).

[5] Inserted.

Life An Old Hag

Life, an old hag
Gathering rags
Night and day in street and alley
Cunning, pitiful, with a mad laugh
Unkempt hair and dirty teeth
Her clothes an empty[1] dark[2]
Forest of rags

See, the wind blasts from her hand[3]
Old paper earrings
And she
Turns over in fury!
More wretched than before
who could bear such deep loss?

Vanquished by the wind, she bends,[4] fatigued
As if finding treasure at her feet
Life, what will you find, peering
Into the well of your past?
What news will you bring
From an empty well's putrid, poisonous air?
At its bottom nothing but pebbles
Nothing but an echo.[5]

---

1 سونا here is an adjective meaning 'deserted'.
2 The word 'faded' is omitted here.
3 Lit. have flown up suddenly.
4 Lit. Now defeated by the wind she has bent down.
5 Lit. voice.

Desire[1] *is a Nun*

Desire is a nun, forlorn,[2] lonesome, sad
Passing her days among
These eternally deprived monks, guardians of the temple
In these palaces of unmoving months and years
On the temple falls a huge shadow
Blood streaking from the face of the worshipped[3]

The nun fumbles out of the temple at night
In her hand a flickering lamp
Staggering, stumbling against walls and doors
In her heart she says, perhaps this lamp's flame
Will be like the brilliance shining far from the temple
The torch of approaching dawn.

Desire is a nun, forlorn and sad
But how should monks know this
Lost in themselves, terrified even[4] of whispers
How could monks know
why the nun is sad?
Standing as they do, like marble slabs
In the soul-burning wilderness of
Endless[5] submission[6]
Where man's passion
Is a wilting rose[7]

---

[1] Also 'hope'.
[2] Also 'helpless', 'destitute'.
[3] Or 'god', 'deity'.
[4] This is inserted to emphasize the sense.
[5] Or 'boundless'.
[6] Or 'impotence', 'humility', 'helplessness'.
[7] Lit. Where the roses of man's feelings do not grow; can also mean 'sorrow' or 'compassion'.

100

The nun wanders, lamp in hand
Stood at the temple door, she thinks its light[8]
Might make the grass glimmer with dew
And over pebbles she might hear
A footstep.

---

[8] Lit. thinking that from this (i.e. from the lamp).

Gazelle[1] *of Night*

O gazelle of night
How shall I slake your thirst
Or show you the mirage in my soul?
That mirage, fearful sorcerer
Who beguiles the heedless[2] traveller
On the road from dawn till night
Child of mirage, mirage-maker
Hovering behind every step
In a thousand ever-changing faces
This despot[3] over uncertain deserts
Has possessed me like a belief[4]
My life filled with his darkness[5]

O gazelle of night, that miscreant lost
In dreams my befores and afters
In veils my nears and fars
How shall I lift the veil
Drawn across my heart's frame
To look inside my soul
Where fear and grief have no claim[6]
And faded is this mirage
O gazelle of night.

---

[1] A term, implying fragility or delicateness, used in Arabic and Persian poetry to address one's beloved.
[2] Or 'simple-hearted'.
[3] Lit. the all-embracing ruler.
[4] Also, 'certainty'.
[5] Lit. He has overcast my life and existence.
[6] Lit. Where there is no trace of fear or grief.

That Lonesome[1] *Word*
(Which desires to unite with Meaning)

Our limbs, raised in prayer toward the sky:
(Don't look at the sky!)
Is this to bypass harm in a fragile place
Or a pretext for hiding from our deprivations?
That ancient[2] and high god (Paradise in truth)
Shall one day deliver us from god
For we are on this earth as a lonesome Word[3]
(But it won't be a world like this) silent and speaking
Which lives in desire for accord with meaning

Grant us the dream of M'arri[4]
(Grant each a smile fitting his vision's breadth)[5]

---

[1] Also 'single' or 'unique'.

[2] Or 'venerable', 'great'.

[3] This is deliberately capitalised so as to imply Word as in λογος.

[4] An eleventh-century Arab thinker and prosodist who was sceptical of God's existence. One of his verses has been translated by Henry Baerlein as follows:

> Now this religion happens to prevail
> Until by that one it is overthrown,–
> Because men dare not live with men alone,
> But always with another fairy-tale.

But although a cynic, Ma'rri's work shows compassion and sometimes faith in his fellow men, as reflected in this verse translated by Ameen Rihani:

> A church, a temple, or a Ka'ba stone,
> Koran, Bible or a martyr's bone–
> All these and more my heart can tolerate
> Since my religion now is Love alone.

(*Anthology of Islamic Literature*, ed James Kritzeck, Meridian, 1975 pp. 126–8).

[5] Lit. Grant each a smile according to the capacity of his own eye.

Grant us the vexation of M'arri's soul
(Where our bold sins[6] will cure all pious misery)
For his dark, unlit eyes[7]
Would pierce
The dark nights of Man's soul[8]
In that world they would see
The chasm between meanings and soul-consuming words
Paradise, for him, was the bliss[9] of simpletons
A refuge from tyrants' whips
The Word was the footprint which might lead them
From the burning sand of estrangement[10]
And in the loneliness of Adam's offspring
In Eve's womb, they might find an
End[11] to their waiting
(Paradise a vast zero; we are the missing term
Without us what equation can hold?[12]
What talk of union between word and meaning?)
Since time began we
Are on this earth as old men
But the heart is still strong
And wise through folly[13]
Grant us the dream of M'arri
(If in Paradise there's joy—and logic—there's grief also

---

[6] Lit. for the courage of sins.

[7] M'arri was orphaned and blinded as a child.

[8] Or 'heart', 'conscience'.

[9] Lit. safety, peace.

[10] Lit. That [the word] was the footprint whereby from the parched sand of estrangement/they might find the opportune day for escape.

[11] Lit. the extreme of waiting.

[12] Lit. can be made.

[13] Lit. confused action.

[14] Lit. uniformity or consistency.

If there's honey, there's poison too)[15]
For we are still in this world a lonesome Word
(Keep your Paradise, show us our equal!)
Which longs to kiss[16] the lips of meaning.

[15] Lit. If there is one jar of honey for all, there is poison.
[16] Lit. unite with.

Wingless

When the dreams of some lost kingdom
Allure, like tears and laughter,[1] the traveller's heart
(Who is wandering like prayers in the dead of night[2]
See, they come and go like mysteries brought to sight)[3]
And the traveller could not hear
Below that laughter and tears, the voice of downtrodden blacks
(We are those on whom Kings have cast their royal gaze
Where the beggars who excel our beggarly plight?)
Who light up rooms in alleys even today;
He could not hear the lash[4] of a tyrant's whip
(Slavery is our task and we the slaves of wealth)
The scant sheet of custom[5] wrapped about our face
(You are god, perform an action as the gods might)
And could not hear the hollowness of ravaged hearts
Where fainting hopes had found refuge
(In the empty night, buildings fill me with dark fear
Fears descend in the heart like some demonic sprite[6]
Why ask the reason for this apathy of ours?
Who laid cruelty's foundation on this earthly site?)
Though every atom[7] of this age lie[8] in the wind's path
And memory be illusion,[9] a protesting of desires
Will the shawl of past luxury slip from its head?

---

[1] Lit. sometimes like tears, sometimes like laughter.

[2] The parenthesized sections of this poem form a ghazal with the rhyme scheme: *aa ba ca da eạ*.

[3] Lit. Like removers of difficulty or like solutions to enigmas.

[4] Lit. sound.

[5] Also 'constitution', 'manner' or 'fashion'.

[6] Lit. Fears enter the heart like evil spirits.

[7] Lit. every thing.

[8] Lit. may be.

[9] Lit. may be an illusion.

The affluent dust of expiring possession?[10]
When the dreams of some lost Kingdom
Allure the traveller's heart
Sometimes like laughter, sometimes like tears
They return, these dreams[11]
In the redness of a tunic
In the desire for chains
As the challenge[12] of love
They return in the dark night of human winglessness
They return
In the sickness of the soul.

---

[10] Lit. night of possession; the implication seems to be that as night comes, possession will fade.

[11] 'These dreams' is inserted to clarify the sense.

[12] Or 'call', 'cry'.

X = Man[1]
(We are Nothing but the Joy of Love)[2]

We remember the far-off Tree[3]
Where we started
Returning to it (eternity is blind)
Again and again
(I am the memory embracing beginning and end
Abode and destination)
We have nested in that same Tree
Seeking it with the same fondness[4]
Always turning back to
That same night's rendezvous

This space-time in which we're a question
Neither a thing nor symbol
But a black mole which
A thorn's point might tear
(I am the event, trapped like thousands
In the soul's snare
Moving, like them, towards
That same night's rendezvous
One flicker of my eyes
Is a chain of accidents
Between which is not a moment's space
I am the memory imprisoning all[5] events)

---

[1] This is the title given to this poem by Ráshed in his own translation of the poem. The parenthesised title is that given to it in the book *X = Man*.

[2] Lit. We are entirely the joy of love.

[3] Presumably, the tree of knowledge of good and evil in the Garden of Eden.

[4] Or 'enthusiasm'. But 'fondness' also encompasses man's original foolishness in eating of the Tree.

[5] Lit. these.

We are nothing but Love's delight
But without a nest, merely a fancy, a thought
Moving together to become one language
A song, a story, speech
Yet fearing, at every step, lest this great search
End in mystery[6]
We know we are the end[7]
The brush, the screen[8]
Sometimes the line, sometimes the feature
But not the painter, nor the painter's masterpiece.

---

[6] Lit. Will become the secret of the search.
[7] Lit. our own end.
[8] Or 'curtain'.

THE POSSIBILITY OF DOUBT

گمان کا ممکن

O Sea...

O sea
The shapes of night, body, voices
Blood running, roaming in veins
Trickling over stones
The shapes of night[1]
Sounded azan[2] as a dancing call
Reaching me,[3] as I lay
Hidden among dying trees
With one hand
On the breast of a bygone age
The other, groping for a path
In a future city:
City, where life's festive colours shall blossom[4]
Where the wine of longing shall be poured!

O sea
Future days have this fear:[5]
When night's disturbed dream[6] withdraws[7]
Into day[8]
What will answer its intricate questions?
The tip of which ray?
The dream of which flowers?

---

[1] This repetition is absent in the original but is inserted to make the sense clear.

[2] Normally the call to prayer.

[3] Lit. And I was listening.

[4] This could also be rendered as: 'Where life will play out its part with colour.'

[5] Or 'anxiety'

[6] Lit. nightmare.

[7] Lit. fades.

[8] Lit. As day breaks forth.

O sea
I shall count
Drop by drop your tears
Where rages the noise of non-being
I shall count your tears, drop by drop
Falling on a future festival
A wedding, garlanded by light
Falling from a future city[9]
I myself invite you to this festival
Where else is there comfort
Except in your waves?

Growling at night on this shore
Pining dogs, moments of sorrow[10]
Casting glances on the moon
Their howls have been thrown far

O sea
How today, the ancient leaves of cloud
Fly on the wings of a stale[11] hope
Bringing from afar a strange story
Sailing in the arms of the moon's broken ship!
The colour of the future city's limbs
—As if all were ready to give their lives—
Has woken in their limbs
Like song
I too have woken![12]

---

[9] Lit. Where for the future of a wedding/The timeless spirit of a future city/Has weaved garlands of rays.

[10] Lit. Pining dogs of sorrowful moments.

[11] Or 'old,' 'ancient'.

[12] Lit. I too have woken in my limbs.

O sea
The waves[13] of tomorrow's festival
Surge within sight of the future city
Unbound by accursed night and day
Men and women will emerge from their homes
Bearing leaves and fruit
Whose touch will turn towards us once again[14]
Spring's vanished face
O sea...

---

[13] Or 'surge'.
[14] Lit. By touching which the vanished Spring will return.

Bid Me Farewell

Bid me farewell
O my Self
What will they say
Those who longed, for a thousand years
For my works?
Bid me farewell
With you I've taken much asylum
In your dark cave
I have tested my limbs to incandescence[1]
In the heart's flame[2]

Bid me farewell
So I may hear the
Silent tears of water and dust
The rustic greeting of Life and Death
Night and Day whispering from their inviolate beyond[3]

Bid me farewell
The late hour[4]
Tolls[5] a twentieth century night
Trees and rocks, beasts and tired birds
Busy for millenia on the earth
Are fixed in conversation
What can they say? Like the gods, I...
Like eternal infidelity[6]

---

[1] Lit. I have heated my limbs/I have caused my limbs to glow. تپانا also refers to the testing of a metal.
[2] Lit. fire.
[3] Lit. The inaccessibility of the limbs of night and day.
[4] Lit. As late as late will be.
[5] Lit. The clock strikes.
[6] Lit. Like eternity's infidelities.

Have turned away
Once again from the pledge of friendship
Bid me farewell, O my Self
You have come to your inlets and seen
In the surveys of the ephemeral mind
Dry leaves scattered by terroristic Autumn:[7]
The cry of non-being's dust
In the city of Being
As if icy calamity[8] had set
In the blood's quagmires
Come to your inlets and see with your own eyes
The dream of mystics on the Eastern horizon
A coffee-coloured dream
Where no hope rests
And no eye mourns
The dying colour and light on the Western horizon

Bid me farewell
But don't descend on your stairways
For they burn in paralysing flame[9]
Bid me farewell, but don't draw breath
Lest the new leaders
Grow fearful at your voice
For you have placed on them forever
The burden of prophethood, their terror!
See there, on the other side of light
Thought turns to paper
Words flee.[10]
The age moves swiftly
Where the Time-players
Are absorbed in a game

---

[7] Lit. Autumn of terror.
[8] Lit. the icy coldness of calamities.
[9] Lit. For they are burning in a fire of senselessness.
[10] A line, superfluous in English, has been omitted here.

Where Life's foundations[11] pray[12]
For rain
There are the parched[13] spaces of doubt

Bid me farewell
For the city gates are open yet
And its people may slide into sleep
Like sand in sacks
O my Self, let me
Escape
So the call, the aspiration, of this tongueless age
Might be heard in every street
Does the new city
(Enveloping hunger's skeleton)
Look to any vision[14] of bread and water? [15]
With hands washed in my own age, I
Shall raise new trees
In their thirsty orchards, I
Shall scrape the layers of coal tar
From their gold and silver—from their body and soul
And I
Will shovel from their doorstep
The piles of snow
Freezing[16] the new city's roads
Bid me farewell
Within myself I have lived
So many dreams; so many wounds
Have I patched myself
I have no strength.

---

[11] Lit. roofs and doors. This has been altered to a more congruous metaphor.
[12] Lit. yearn.
[13] Lit. thieving. Again, the metaphor has been adapted.
[14] Lit. جھلک means 'glimpse', 'reflection' or 'refulgence'.
[15] The question form is inverted here.
[16] Lit. blocking.

Speak

Speak to me
Show me your face, for mine
Is scorched by the fire[1] of your eyes
Speak to me
Draw from my face the veil which bears
A rainbow coloured by hypocrisy
A rainbow on which
No reddening dawn can descend[2]

Yesterday you saw
A beggar (it was I)[3] benumbed[4]
Walled within the shadow of a morning sun[5]
Your lips agape, your eyes staring
How could I know for sure their warmth
Deep as I was in my heart's calamity[6]
Poeticising with memories

Speak to me
For now the space between night and dawn
Expires.[7] Speak to me, for your words
Will shroud[8] the face of Death
Descend now beyond eye and ear and lip

---

[1] Lit. heat.
[2] Lit. where dawn's zeal does not descend.
[3] In the original the parentheses enclose the word 'beggar'.
[4] Lit. You saw that...I was found benumbed.
[5] Lit. Beneath a shadow cast by the wall of morning.
[6] Lit. For I was in the depth of my heart's calamities.
[7] Lit. Vanishes.
[8] Lit. wipe out.

118

Onto the highways of ruined cities
Light the candles of sound
So the ripples of mystery might surface
Circle within circle.[9]

---

[9] Lit. line within line.

The Possibility of Doubt:[1] *Which is You and I*

Bountiful sun
Giving its roundness to a cold stone
Giving its flatness
(That stone rolls, like me, in sapless[2] greenery
In memories of an age of sand and wind)
Its bounty intoxicating[3]
The meandering waters
It doesn't know me
But perhaps it is I
Shrouded in the illusion
That I am my own proof and answer;
It cannot know
My soft voice, floating
From the far[4] shore of time's lake

Here is the lake on whose surface
A myriad mortals[5]
Walk parallel to the horizon
Borne across
By rippling time[6]
But those intent on a skyward journey
Are made to hear the sound
Of infernal songs[7]
Drowned in time-waves!

---

[1] گمان can also be rendered as 'suspicion'.
[2] Lit. brown.
[3] Lit. the intoxication/happiness of its bounty.
[4] Lit. other.
[5] Lit. men or human beings. 'Mortals' seems a more appropriate rendering given the cosmic scale of reference of this poem.
[6] Lit. Time's waves.
[7] Or 'the songs of hell'.

All sailors fear this sound
For in the lake sits a secret thief
Of skywardness[8]
His locks hanging from the horizon's roof
Crying: 'Come now, come!
I am time, awaiting you
I know the secrets of domes
Tree, minaret, tower, stairway
All are my companions
I am the final rest of every aircraft
The defining shore of ships' seas
Come now, come!
Endless tales like yours I have borne
To the bosom of their eternity'.
All sailors evade its voice
On the horizon's worn highways all move in fear
But those skybound
Gathered at the threshold of high and low
Are descending by this vertical rope[9]
The rope by which they rise

And I have travelled with them
Reaching the shore
Where a god's footprint has taken refuge
Where his frail eyes
Yet survive
This heavenward journey, my hope[10]
Is now my fate

But why
Did I choose a path

---

[8] Lit. a thief of perpendicularity secretly sits.
[9] Lit. rope of the perpendicular.
[10] Lit. which was my longing.

Beyond any union with Self?[11]
A path ending
On a shore of the heart's mystery?
Before the shore lurks
The possibility of doubt
Which is you, which is I

In the hope of finding you (myself)
I came
In search of a possibility
Which is you, which is I
A face
Which is you and I
Wandering after[12] an image
Which is you and I

In this search
How many beginnings have I counted
(I fear him who says:
I have no fear)
In this search how many alleys
How many crossroads
How many silent statues
Have I passed
What[13] gardens[14]
What drunken nights
What embraces
The seas of what loves
Have I passed
What lanterns of sense and action

---

[11] Lit. Where there was no possibility of union...
[12] Lit. in pursuit of.
[13] Lit. How many.
[14] The repetition of 'In this search' is omitted here.

The domes of how many faiths?
How many beginnings and endings
Have I counted
In this search now there is no door
No unborn age
Each remaining terminus[15]
Is a passing tale
All roads, all mastered questions, have lost meaning[16]
All answers assume the face of history
Repeating only themselves—
'We are the answer, we are the answer!
We know for sure we are the answer!'
How certainly they echo[17] certainty!
But they all lie in contradiction
Like the possibility of doubt
Which is you and I

All the logs (you well know)
Which flow on the river's surface
Know this to be an accident
With no refuge[18] for them (for any)
Logs floating on the river's surface
To become crocodiles is not their fate
(A crocodile's beginning includes its end[19]
A crocodile's heart a crocodile's heart!)
Nor is their fate to grow as trees again
(A tree's beginning includes its end[20]
A tree's heart a tree's heart!)

---

[15] Or 'destination'.
[16] Lit. importance.
[17] Lit. repeat.
[18] Lit. From which there is no refuge...
[19] Lit. includes a crocodile.
[20] Lit. includes a tree.

Before the poor logs, all backward paths
Are closed
On the river they swim with the river's force
Now their end is a harbour which
Ever welcomes[21]
Now their end is a ship
Beyond the builder's conception
Now their end
The leaves on which dark[22] words will print
Books for whom there is no reader
Artists' canvasses
With as yet no face
On which might fall the shade[23] of tears
Whose still unpictured vision
The future might fill[24]
Before the poor logs, all backward paths
Lie closed
Those streams of imagined survival, open till now
Stand before them as the possibility of doubt
A possibility[25] which is you and I
You and I.

---

[21] Lit. Ever holds out its arms.
[22] Or 'evil'.
[23] Lit. colour.
[24] Lit. In which the future/Could fill the picture of their vision.
[25] 'of doubt' is omitted here.